Artificial Intelligence Mastery

4 Books in 1:

Machine Learning and Artificial Intelligence for beginners+AI for Business+AI Superpowers and Data Analytics+IOT, Data Science and DL updated edition

Jeff Mc Frockman

Table of Contents

Artificial Intelligence for Beginners

Machine Learning for Beginners

Artificial Intelligence and Machine Learning

Artificial Intelligence Business

Artificial Intelligence for Beginners

Easy to Understand Guide of AI, Data Science, and Internet of Things. How to Use AI in Practice? Revelations of AI Superpowers Explained for the Real World

Jeff Mc Frockman

Introduction

Congratulations on downloading *Artificial Intelligence for Beginners: Easy to Understand Guide of AI, Data Science, and Internet of Things. How to Use AI in Practice? Revelations of AI Superpowers Explained for the Real World,* and thank you for doing so. With the constant increase in the generation of big data, we are forced to look for technologies that can help us make informed decisions. All professional fields have become very complex due to the growing demand to find the most efficient ways to extract value from the existing data. By downloading this book, you have taken the first step towards learning how to use big data and how artificial intelligence can make it possible to utilize the available data productively. The information that you find in the following chapters is very important as it will help you understand the role of artificial intelligence in society and its impact on your career.

To that end, this book provides an in-depth overview of artificial intelligence, highlighting its major concepts, its historical development, and application in various fields, including finance, business, and medicine, and the types of programming that are significant to artificial intelligence. It also covers the use of robotics and their role in the advancement of artificial intelligence, as well as how each influences the other. We have also comprehensively addressed the role of the Internet of Things (IoT) today. People can utilize so much from artificial intelligence and IoT to improve their lives and enhance their productivity. An interesting concept that we have also covered in the AI superpowers across the world, with a particular focus on China, which is the fastest rising region for AI development.

There are several books on Artificial Intelligence in the market, thanks again for choosing this one! I hope you enjoy reading!

Chapter 1: Artificial Intelligence Basic Terminologies

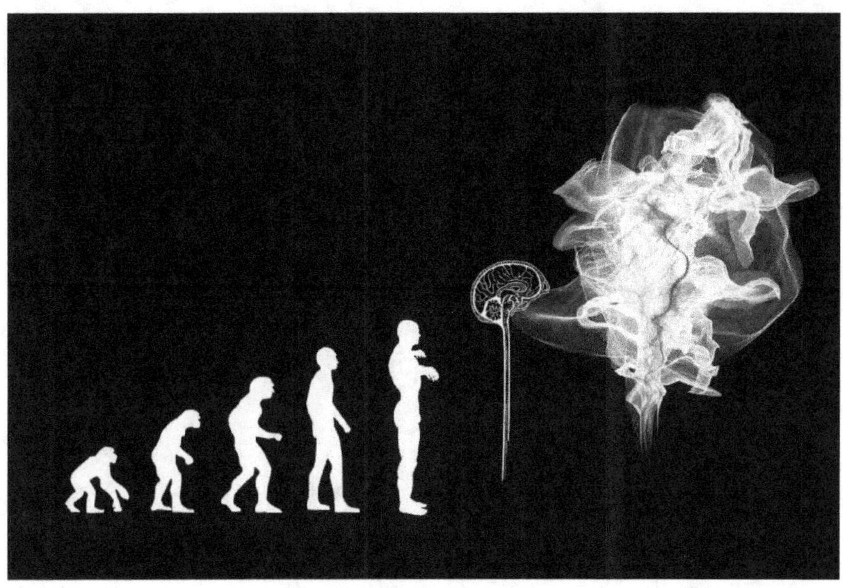

Artificial intelligence or simply AI is a term used in computer science to refer to the intelligence shown by machines as compared to the natural intelligence which human beings possess. The machines copy the human mind's functions and capabilities. The field of AI cut across key science areas such as human psychology, computer science, and cognitive science. There are key terms that are important if you are interested in learning Artificial intelligence. They include the following:

- **Abductive logic programming**

  This is one of the terms which you need to be familiar with. It refers to the area of AI that deals with the representation of information about the general world in a way that a computer system can make use in finding solutions for difficult tasks. The problems are arranged in an orderly manner to allow the computer to reach a conclusion based on a given premise.

- **AI-complete**

In computer science, problems can be classified either as AI hard or AI-complete. An AI-complete or AI hard problem means that a problem can be as difficult as that of getting a solution to a key artificial intelligence. This then makes computers to be comparable to humans in terms of intelligence.

- **Approximate string matching**

This is a method of locating strings that exactly match a given pattern. Approximate matching involves matching the strings by dividing it into two subproblems. You then locate the substring, which shares the same matches within the given string. String matching is an important cluster of string algorithms which are used to locate where one or more of strings are located inside a bigger text or string.

- **Algorithms**

Algorithms basically refer to the set of rules or instructions which are given to an AI or other machines for the purpose of enabling it to learn on its own. Some of the key kinds of algorithms include classification, clustering, recommendation, and regression.

- **Artificial intelligence**

  This refers to the machine's ability to make significant decisions, which are similar to those made by the human mind. They use the same process as that used by the human mind in making decisions.

- **Artificial neural network (ANN)**

  ANN refers to a model that was designed to behave the same way as a human brain. It generates solutions to problems that prove challenging for normal computer systems to solve.

- **Automatic computing**

  This is a system's capability to manage its own resources in order to function optimally without your input.

- **Chatbots**

  A Chatbot is a chat robot that is made to carry out intelligent conversations with humans. Such conversations are carried out through text chats or voice commands. They make use of a computer interface with AI capabilities.

- **Classification**

  This type of algorithm enables the machines to give a type of specific data using the available training data.

- **Cluster analysis**

  This refers to the kind of unsupervised learning commonly designed for exploratory data analysis to locate missing patterns or groupings of data. They are created in the same way the metrics of Euclidean are.

- **Clustering**

  Clustering algorithms enable the machines to group data points into categories, which share the same characteristics.

- **Cognitive computing**

  This refers to the computerized model which copy the manner in which the human mind thinks. Cognitive computing happens when the machine is taken through a self-learning process through the use of data mining, pattern recognition, and NPL.

- **Convolutional neural network (CNN)**

  This refers to the kind of neural networks that locate and sensibly describes images.

- **Deep learning**

  This refers to the ability of an AI machine to mimic human thought patterns autonomously through neural networks that are made of cascading layers of data.

- **Decision tree**

  Decision tree refers to a decision model used to analyze decisions and the possible consequences of the decisions. The format of the decision tree is similar to a flow chart.

- **Game AI**

  This refers to a type of AI designed specifically for gaming. It makes use of an algorithm in order to replace randomness. Game AI is a digital behavior applied in non-player characters to give intelligence similar to those of humans and player actions taken by the player as a result of their reactions.

- **Genetic algorithm**

  This refers to an algorithm with similar principles guiding natural selection and genetics. It is used to come up with optimal solutions to challenging problems, which could take a long time to solve in normal circumstances.

- **Heuristic search methods**

  Refers to the system support, which reduces the search for solutions to a problem by getting rid of options which are not correct.

- **Knowledge Engineering**

  This is software engineering that generates general knowledge-based systems. The system is generally based and may have content touching on all areas such as technical, scientific, and social aspects.

- **Logic programming**

  Refers to programming type whereby you carry out computations drawn from the available knowledge section of facts and rules. Some of the examples of computer languages using logic programming are LISP and Prolog programming languages, which are widely used in AI programming.

- **Machine intelligence**

  This refers to the general concept, which includes machine learning, deep learning, and the learning of the general AI algorithms.

- **Machine learning**

  This refers to the type of AI which specializes in algorithms. In machine learning, the goal is to design machines that learn without the need for programming. The machines could also be altered when new data is exposed to it.

- **Machine perception**

  The capability of a machine for a system to get data from outside sources. The machine can then analyze and interpret the data in the same way that the human mind does. This is possible when both hardware and software data are attached.

- **Natural Language Processing**.

  This refers to the program which is designed in such a way that it can recognize and understand human communication.

- **Recurrent neural network (RNN)**

  This refers to the kind of designed program that makes sense of sequential information. The program then identifies patterns and create results based on the calculations of the given data.

- **Supervised learning**

  Refers to the learning in which output data sets are trained to give expected algorithms in the same manner in which a classroom teacher trains a student

- **Swarm behavior**

  This refers to the ability of a machine to follow the rules without involving central coordination.

- **Unsupervised learning**

  This refers to the machine learning algorithm, which is used to come up with inferences from datasets without labeled responses from the input data.

The Basic Concepts You Need to Be Aware in Artificial Intelligence

In order for you to understand the more profound concepts which are key in AI. you need to understand some basic terms of AI. These includes:

- **Machine Learning**

  Machine learning is a type of AI that enables machines to learn tasks without the preexisting code. Machines are given a large number of trial tasks. They are then allowed to go through the trials. The machines are then allowed to learn and adapt their own methods in order to achieve the given goals. For example, you can give a machine several images to analyze and recognize. The machine will go through several permutations in order to acquire the ability to correctly identify patterns or shapes in faces.

- **Deep Learning**

  This refers to the generation of general-purpose learning algorithms which aid the machine to learn several tasks. A very good example of deep learning can be drawn from Google's AlphaGo project. The AlphaGo once won a complex game against a professional Go player. This was widely thought to be an impossible task before because of the complexity of the game involved.

- **Neural Networks**

Deep learning occurs when neural networks mimic neurons or brain cells. Artificial neural networks were designed from a biological background. The models make use of the computer science rules to imitate the processes of your human mind, thereby making general learning possible. An artificial neural network attempts to imitate the processes of a highly interconnected human brain cells.

Neural networks are made up of three key layers, which include the input layer, as well as the hidden layer, and lastly, an output layer. These layers carry a big number of nodes. When an input layer receives information and is given the required weight, the interconnected nodes then multiply the given weight of the connection.

Once the unit of information has reached the desired level, it passes to the next layer. machines then make comparisons of outputs from neural networks to learn from past experiences. They then make changes affecting connections and weights arising from their differences.

These three key AI terms make it possible for the hardware and the software robots to acquire unique thinking and functions, which are outside the set of codes. Once you have understood these terms, you can then move to a more advanced AI field, which includes the artificial superintelligence, artificial narrow intelligence, and artificial general intelligence.

- **Artificial Intelligence Areas**

For you to understand the implications of AI on your life and the general society, you need to distinguish the broad types of AI.

- **Artificial Narrow Intelligence (ANI)**

It is also referred to as the weak AI. It is the type of AI that you are likely to encounter in your world today. This type of AI is programmed to perform single tasks, for example, to check the weather conditions, or play a game like chess. Narrow intelligence can perform real-time tasks once they get information from a specific data set. However, the systems cannot be subjected to perform more than one task at a given time.

In addition, narrow AI is not like human beings in the sense that they lack emotions associated with humans. They are not also conscious or sentient. Narrow AI can only function within a preset range even though it will appear to you to be complex than that.

It is much probable that all the machines you see today are narrow AI machines, a good example is, Siri the popular Google Assistant.

The reason it is called weak is that they are can never be compared to human intelligence. They lack the original intelligence comparable to that of human intelligence.

Although it is referred to as weak intelligence, this should not be a reason for you to take it for granted.

ANI systems are designed to process data and very quickly complete the assigned tasks then a normal human being. In this regard, ANI systems have been able to make improvements in the productivity of humans as well as in the quality of life that human beings enjoy.

ANI has also enabled you to get rid of the boring, monotonous tasks that are not enjoyable for you. It has made your lives significantly better. For example, you are now able to escape the frustrations that come with long traffic jams courtesy of self-driving cars or automatic cars. In addition, From the currently available ANI systems, more advanced AI systems are being developed.

- **Artificial General Intelligence**

Another name for AGI is strong intelligence. AGI refers to machines that have the same intelligence as that of a human brain. They can do any function of your brain. This is the sort of AI that is found in sci-fi movies. The operating systems of AGI machines are not only conscious, but they are also sentient. AGI machines also possess emotional and self-awareness.

They can process data faster than any human being. AGI is able to make concrete reasoning and can conclusively solve any given problem. They can also make clear judgments. In this regard, ANI is considered to be very good at innovation as well as imagination.

- **Artificial Superintelligence**

  Artificial superintelligence is expected to surpass all human abilities in all areas, including problem solving, creativity, and wisdom. These machines will possess great intelligence never recorded before in human history. It is a type of AI that has got many people worried about the possibility of the human race getting extinct.

How Artificial Intelligence Was Initiated

Scientists have been hard at work for a while now. They were inspired at coming up with machines that have the same intelligence as that of human beings.

The journey started in 1936 when the British mathematician put into practice his theories with the aim of proving that a machine called the Turing machine had the ability to perform tasks similar to those performed by the human brain. He made this possible by breaking down the steps and reprinting them in an algorithm. This was the solid foundation for the establishment of artificial intelligence.

In 1956, a group of scientists came together for a conference at Dartmouth College in New Hampshire. They shared a common agreement that it was possible to learn the key functionalities of the human mind and imitate them into machines. The machine was then given the name, artificial intelligence, by John McCarthy. It was also during this same conference that the scientists developed the first AI program known as a Logic theorist. The program is able to prove several mathematical theorems as well as data.

In 1966, Joseph Weizenbaum came up with a computer program capable of carrying out communication with human beings. He called it ELIZA. ELIZA made use of scripts to imitate various conversations with their counterparts and humans.

In 1972, artificial intelligence entered into the medical field, with MYCIN. This was a system, which was useful in the diagnosis and treatment of various illnesses. They are useful in the diagnosis of diseases as well as in giving out treatment medicine to the patients.

In 1986, the computer acquired the ability to speak. This was made possible by two scientists, namely Terrence J Sejnowski and Charles Rosenberg. They put sample words and phonetical chains to the program, which they then trained to identify the words and correctly pronounced them.

In 1997, an AI chess computer named Deep Blue developed by the IBM Company was able to compete and defeat a reigning world champion. This was widely viewed as a huge success in the field of AI.

In 2011, artificial intelligence made progress into our daily lives. People started interacting extensively with AI-enabled features found in their devices, such as smartphones. The world also witnessed the development of Powerful processors commonly found in your computers or smartphones.

It was in 2011 when a computer program known as Watson competed with human beings in a quiz aired on television. The program surprised many when it won against human competitors. This proved that the computer has the ability to recognize languages and even answer questions in record speed.

And recently, in 2018, the world witnessed a debate between an AI and two master debaters on the complex topic of space travel. It was also in the same year when an AI appointment with a hairdresser. For the entire period, the call lasted, the hairdresser on the other end failed to notice that she was not talking to a human but to a machine.

Advantages and Disadvantages of Using Artificial Intelligence

The advantages of AI include:

- AI Makes minimal errors when compared to the errors that human beings make when subjected to perform the same task. The machines are also able to undertake given tasks fast and accurately.

- Another advantage of AI is that it performs tasks, which are located in hazardous environments. They can perform tasks otherwise viewed as too dangerous for humans to undertake. Such tasks could cause severe injuries or even death to humans.

- They predict the functions you want to be undertaken, for example, when you are typing or searching on your device. They can also guide you on selecting various actions. This way they act as your assistant

- It can be used to detect fraud in card-based systems and as well as possible fraud in other systems.

- AI machines can be useful in your entertainment. Robots are used to enhance various human activities. Some perform entertainment tasks very well.

Disadvantages of AI include:

- It requires a lot of capital to research and comes up with a new AI machine. It also takes a lot of time from the initial stage to launching of an AI machine.

- Some key sectors of the world, such as the human rights groups, have raised pertinent issues touching on AI. They view the advancement of AI as an attempt by humans at creating a fellow human, which is questionable and unethical.

- Access and retrieval of data in AI may not lead to connections in memory like the way human brains function. They cannot also work outside of what they were programmed to do.

- Robots have also been known to replace the jobs done by men. This will lead to massive job losses and will result in economic meltdowns across the globe.

Chapter 2: Robotics

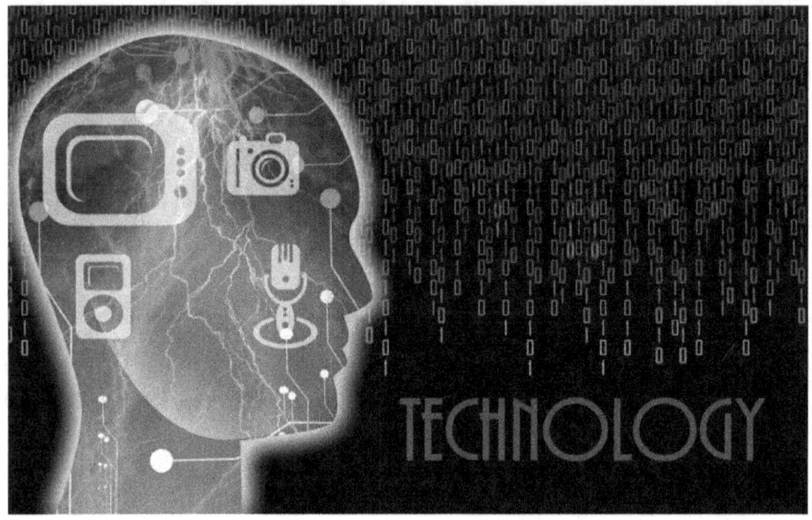

The field of AI has dramatically developed with many new general technological achievements. One advancement is that the rise of Big Data, which offers many possibilities to build programming capability into robotic systems. Another is that the use of the latest varieties of sensors and connected devices to watch environmental aspects like temperature, motion, light, atmospheric pressure, and more. All of these serve AI and, therefore, the generation of a lot of complicated and sophisticated robots for several uses, as well as health, manufacturing, agriculture, safety, and human assist.

The field of robotics additionally intersects with problems around AI. Since robots are physically distinct units, they appear to have their own intelligence, albeit one restricted by their programming and capabilities. This idea has sparked new debates over ancient fictional theories, like Asimov's three laws of AI, that address the interaction of humans with robots in some mechanized future.

Artificial intelligence (AI) is arguably the first exciting field in robotics. It is the foremost controversial. Everyone agrees that a robot would work well in an assembly line; however, there is not any agreement on whether or not a robot will ever be intelligent. Robots are software-driven machines that are sometimes capable of performing a series of actions autonomously or semi-autonomously.

Three vital factors represent a robot:

1. They are programmable.
2. They act with the physical world via actuators and sensors.
3. They are sometimes autonomous or semi-autonomous.

It is said that robots are "usually" autonomous as a result of some robots are not. Telerobots, as an example, is entirely controlled by a human operator; however, robotics continues to be classified as a subsection of AI. This representation can be one instance where the meaning of AI is not entirely clear.

It is tough to get specialists to agree precisely what makes up a robot. Some individuals maintain that a programmable machine should be able to "think" and form decisions. There, however, is no standard definition of "robot thinking." Requiring a tool to "think" infers some level of AI is applied.

Robotics involves designing, programming, and building of actual robots. A minor component of it involves AI. Most AI programs are not applied to create or manage robots. Even once AI is employed to manage robots, machine learning algorithms are solely a section of the more extensive robotic system that additionally includes non-AI programming, actuators, and sensors.

AI involves some level of machine learning, but not always. An example is where innovative design is "trained" to retort to a selected input in a very sure means by use of far-famed inputs and outputs. The critical facet that differentiates AI from a lot of typical programming is that the word "intelligence." Programs that do not involve AI merely perform an outlined sequence of directions. AI programs are built to mimic some degree of human intelligence.

Robots Features

Artificially intelligent robots are the link between AI and robotics. These are machines that are controlled by AI programs. Up until recently, most robots have been programmed to solely perform a series of monotonous tasks. As earlier stated, dull, monotonous activities do not need AI.

Robots need an energy supply, and several factors move into deciding that the style of power provides the foremost freedom and capability for a robotic body. There are many various ways to get, transmit, and store energy. Generators, batteries, and fuel cells provide the power that is regionally kept, however more temporary, whereas tethering to an influence supply naturally curbs the device's independence and variety of functions.

The innovation that empowers automation sense has fostered our ability to speak to machines electronically for several years. Transmission mechanisms, like microphones and cameras, facilitate the transmission of sensory information to computers inside simulated nervous systems. A sense is helpful if not essential to robots' interaction with live, natural phenomena. As the human sensory system is attenuated into vision, hearing, touch, smell, and style- all have already or are in the process of being enforced into robotic technology somehow.

Featured applications of robots include:

- **Computer Vision**

  One obvious application of AI to AI is in computer vision. Computer vision permits robots and drones to explore the physical world much more accurately. This is a technology of AI that the robots use to see. Computer vision plays an important role in the domains of safety, agriculture, health, biometrics, and entertainment.

  Computer vision mechanically extracts, analyzes, and comprehends valuable data from one image or an array of pictures. This method involves the development of algorithms to achieve automatic visual understanding.

- **Unsupervised Machine Learning**

  Robots are already utilized in manufacturing, however, typically in preprogrammed tasks. Robots may learn tasks with machine learning by being taught by humans or through unsupervised machine learning. While there is a concern that robots like these may replace individuals in industrial jobs, these robots may work alongside humans as "cobots", involving more collaboration with individuals rather than taking up their positions.

Some new robots even can learn in minimal capability. Learning robots acknowledge if a particular action (moving its legs in a specific manner, for instance) achieved the desired result (navigating obstacles). The mechanism stores this data and tries the productive action the subsequent time it encounters an identical scenario. Some robots will learn by mimicking human responses. In Japan, engineers have taught a robot to dance by demonstrating the moves themselves

Intelligence, deftness, sense, and power all converge to create self-governance, that successively may, on paper, cause a virtually personified individualization of mechanical bodies. Derived from its origin inside a piece of speculative fictional tale, the word "robot" has nearly universally observed by artificial means intelligent machinery with a certain degree of humanity to its style and thought (however distant). Therefore, robots are mechanically imbued with a way of individuality. It conjointly raises several potential queries on whether or not or not a machine will ever incredibly "awaken" and become aware, and by extension, treated as a person (or personal subject).

- **Human Error**

Another primary application of AI to robotics that has gotten attention in recent years is autonomous or self-driving cars. This sort of use is enticing because it promises to reduce human driver error that is the cause of most traffic accidents. A robotic automobile will not get tired, impaired, or inattentive while the human driver will. Even though there are several high-profile accidents involving autonomous vehicles, they show plenty of promise to be considered safer than human-driven cars.

A significant area of analysis involving robots and AI is in medical technologies. Robots within the future might perform surgery without intervention from a doctor. Like autonomous vehicles, robotic surgeons might perform delicate operations for extended periods than human doctors will, while not feeling tired or making mistakes.

- **Dexterity**

Dexterity is the practicality of organs, limbs, and extremities, likewise because of the general varies of motor ability and physical capability of an animated body. In robotics, quickness is maximized wherever there is harmony between high-level programming and subtle hardware that comes with environmental sensing capability. Several alternative companies are achieving important milestones in robotic quickness and physical interactivity. This technology application lends an excellent deal of insight into the longer term of robot quickness. However, not all robots mimic the human physical type (those that do are usually mentioned as "androids," whose Greek chronicle origin essentially interprets as "likeness to man").

Classification of Robots

The most popular robot classification includes immobile and mobile robots. These two types have entirely different operating systems and thus have different capabilities. A majority of **Fixed robots** are industrial robot operators who work in well-outlined environments tailored for programmable machines. Industrial robots perform specific and dull tasks such as bonding or painting elements in automotive manufacturing factories. With the development of human-robot interaction devices and sensors, robot operators are more and more used in a minimally controlled setting like surgery, which requires high precision.

In comparison, **mobile robots** maneuver around and carry out multiple tasks in considerably vast, ill-defined, and unforeseeable environments that do not seem to be designed explicitly for robots. These robots have to modify things that do not seem to be precisely renowned but which change over time. Environments like these will embody unpredictable entities like humans and animals. Some of the common mobile robots include robotic gutter cleaners and automated self-driving vehicles.

There lacks a clear distinction between the functions meted out by mobile robots and immobile robots. There exist three primary **environments** for which mobile robots would need considerably variable class principles as a result of the difference in the means of **motion**: *terrestrial* (for instance, cars), *aquatic* (for instance, underwater expedition), and *aerial* (for instance, drones). The classification is not strict; take, for instance, some amphibious robots that move on water and the ground. Robots that operate in these three terrains are further subdivided into pseudo groups: terrestrial robots either have legs, or wheels, and aerial drones are light balloons or heavy craft, that are successively sub-grouped into fixed-wing and rotary-wing as in the case of helicopters.

Nowadays, robots do plenty of various tasks in several industries, and therefore the variety of jobs entrusted to robots is growing steadily. Robots can also be grouped in accordance with the supposed **application** industry and the functions they perform as follows:

- **Industrial robots** that perform repetitive duties on manufacturing tasks are mentioned. Industrial robots are robots employed in industrial manufacturing surroundings. Sometimes these are articulated arms developed explicitly for applications such as assembling, product handling, painting et al. If we tend to decide strictly by uses of this kind, we might additionally embrace some automatic guided vehicles and different robots. The first robots are said to have been industrial robots as a result of the well-defined surroundings simplified their style.

- **Service robots**, alternatively, assist humans in their tasks. Domestic or social robots include several quite wholly different devices like robotic sweepers, vacuum cleaners, gutter cleaners, robotic pool cleaners, and various robots that may do completely different chores. To add, defense applications like intelligence activity drones and telepresence robots can be thought to be home robots if employed in that environment. Service robots do not constitute different varieties by usage. These can also be completely different information-gathering robots, robots created to indicate off technologies, robots used for analysis, etc.

  Robots have also been increasingly used in the medical field, in surgeries, training ang, and rehabilitation. These are examples of applications that need sharper sensors and better user interaction. **Medical robots** are employed in drugs and medical establishments. The very first medical application being surgery robots.

- **Military robots** are employed in the military. These kinds of robots include explosive diffusion robots, transportation robots, and intelligence activity drones. Usually, robots created at first for military functions will be employed in law enforcement, search and rescue efforts, and different connected fields.

- **Entertainment robots** used for recreation. This is often an inclusive class. It starts with toy bots like Robosapien or the running grandfather clock and culminates with real heavyweights like articulated robot arms used as movement simulators. Hobby robots are also in this class. They constitute those that you create for the sake of code. Line tracker robots, sumo-bots, are robots created only for fun and competition purposes.

- **Space robots** would come with robots used on the International Space Station. Mars rovers and different robots employed in space exploration.

The Three Laws of Robotics

In fiction, the three laws of robotics are a collection of 3 rules written by Isaac Asimov, introduced in his short work of fiction in 1942 the rules go as follows:

1. "A robot may not injure a person or, through inaction, permit a person to come to harm."

2. "A robot should adapt orders passed by persons except when the aforementioned orders would conflict with the First Law."

3. "A robot should defend its own existence as long in and of itself protection does not conflict with the primary or Second Law."

Asimov's laws are an endeavor to handle the AI-uprising threat. The technical obstacle in creating robots abide by them is our current limitation in making them observe the commandments. The major obstacle, a philosophical and moral one, is our assumption that given such imprecise constraints, the robots can behave specifically however we would like them to, even though we do not understand what we tend to mean.

Chapter 3: Software and Programming to Make Predictions and Intelligent Decisions

Artificial intelligence has made substantial progress into every part of your life, and this is projected to continue in the near future. AI has made possible for automation of tasks for you as a user, which has enabled you to have a unique experience as you interact with your normal devices such as the smartphone. Automating these processes saves you of your time and energy and makes your job easier allowing you or your employees to work more efficiently and productively.

AI software provides developers with tools to build intelligent applications. The tools include algorithms, and libraries, or frameworks of code, which is useful in the creation of important functionalities for the software. AI software enables you to be more productive, enabling you to carry out otherwise boring and monotonous tasks using your AI machine. It also helps companies in coming up with key decisions.

AI software is also used by software professionals to come up with solutions which are beneficial to workers in all kind of professions. Software applications are important when generating new applications. They can also help you to improve on any existing software application. Moreover, you can also use AI software to carry out general machine learning capabilities or deep learning capabilities

Importance of AI Software

Some of the significance of AI software are:

- **Useful in creating conversational interfaces**

  Various Software companies are striving to improve and compete with the latest development in AI in order to come up with competitive AI products like Amazon's Alexa or Google Home a use AI software.

- **AI software helps in personalization**

  You can create a high level of personalization by using machine learning algorithms. Personalization improves your software products for all users; thus, you offer them a unique experience. Companies that have successfully used AI in personalization includes Amazon, which uses the software to personalize their consumer shopping and Netflix in their movie recommendation capabilities.

- **AI software useful in intelligent decision making**

AI helps humans to make logical assumptions instead of the other way round. Machine learning is important for you when making decisions touching on your business because it provides you with evidence and predicted consequences of the decision you are making. You are enabled then to avoid costly human errors. It can also help equip users with the information necessary to defend the decisions they make.

- **Predictive capabilities**

The predictive feature provides outcomes which the platforms take to be right. For example, in business, you can use the software to have expense management applications add an expense to you repost on its own. You can easily insert this kind of application with your AI software.

- **AI software is useful in the automation of monotonous tasks**

You can use machine learning in automating challenging tasks that you do on a daily basis in your workplace. By using AI for these tasks, you save a lot of time, which you can use to carry out other more productive tasks. Of importance is the fact that contrary to popular beliefs, AI use does not replace human jobs, but it serves to complement them.

Artificial Intelligence Software Types

The types of Artificial intelligence is quite wide. There are key categories of AI, such as chatbot, AI platforms, deep learning, and machine learning, which you should know.

- **Deep learning**

  Deep learning makes predictions and decisions using a neural network. With artificial neural networks, important decisions are made in the same way in which the human mind normally makes decisions.

- **Machine learning**

  This algorithm category consists of a wide variety of frameworks performing various functions from the available data. They, however, require some element of personal training. You require enough training and experience to incubate an intelligent application using a machine learning algorithm.

- **Chatbots**

  Chatbots are an advance area of AI learning. They are created to perform a specific function, for example, for automation in business and to help in creating a unique customer experience. The applications interact with customers using voice or texts.

  Chatbots can also be used in call centers by agents. It can also be used to carry out live chats with potential customers. Businesses can also determine customer need by using chatbots.

Moreover, chatbots are often used as tools for customer support or virtual assistance. Chatbots can interact with customers over a long time. This way, as they interact, they also learn new vocabulary and intelligence.

- **AI platforms**

AI platforms give you good solutions, especially if you are trying to build your applications using another platform. These tools give you a drag and drop option to assist you in building the app from zero. The platforms equip applications with an intelligent advantage; they possibly make the creation of smart applications less costly and fast. However, you will need to be highly skilled and knowledgeable to maneuver through the platforms.

- **AI programming for the beginners**

Artificial intelligence refers to the branch computer science that involves the development of machines with similar intelligence as that of humans. Artificial intelligence experts were interested in coming up with systems that are unique and which could surpass human intelligence. The advancement in technology has made it possible to create self-aware robots, which will become part of your life in the coming days.

The Goals of AI

Some of the goals of AI are specific. This includes reasoning, planning, and scheduling, natural language processing (NLP), among others.

- **Knowledge and reasoning**— The main goal of AI is to focus on the implementation and designing of computer representations, with the aim of processing information. AI automates different types of reasoning through the use of codification of relationships so that they can be interpreted easily by the computer system. Knowledge representation and reasoning are usually applied in a natural-language user interface that facilitates communication between computers and humans.

- **Automated planning and scheduling**—AI is also concerned with the generation of automated action sequences, which corresponds to AI systems measurements. The goal of AI is to mechanize and automate the generation of planned actions through planning and scheduling. One of the examples of AI planning is the self-correcting robots that act as computerized suggestions.

- **Natural language processing (NLP)**— Another goal of AI is to process natural language. NLP entails an analysis and generation of language that can be used for computer interfacing. The NLP process aims to implement specific computer systems that can be used to process big data.

- **Computer vision—** The main aim of the AI component is to assess the automation and computerization of activities, which can be performed by system development to process and interpret visual data. Computer vision utilizes several applications such as object recognition, video tracking, and automated image manipulation. Computer vision is commonly applied in automated image manipulation, object recognition, virtual reality integration, and video tracking.

- **Robotics—**Another significant goal of AI is to construct, design, and operate robots and machines that can replace human tasks. Robotics refers to the scientific branch that entails electronic engineering, computer science, information engineering, and electronic engineering. Research on robotics is currently conducted to promote military, domestic, and commercial applications.

Through these goals, AI has greatly influenced a wide variety of sectors, including education, transport, security, infrastructure, and communication.

Classification of AI

There are three common categories of AI-based on their capabilities:

- **Weak Artificial intelligence**

  Weak AI is also known as Narrow AI. It was designed to undertake a single task. You cannot find any real intelligence in weak AI. You can't also find any self-awareness with weak AI. A good example of a weak AI is iOS Siri.

- **Strong AI**

  It is also referred to as True AI. The software performs like the human brain. It undertakes tasks which a normal human can do. A good example of a strong AI is the Matrix I Robot.

- **Artificial superintelligence**

  This is an intellect that can perform better than the best human brains in all areas, including scientific creativity and social skills. Because of advancements in artificial superintelligence, many people have raised concerned about the possibility of human beings becoming extinct and being replaced by super-intelligent robots.

How to Get Started as an AI Beginner

As a beginner in understanding how AI works, it is essential that you follow the steps below:

- **Step 1: Learn a programming language**

  Learning the programming languages is one of the first steps you need to undertake. It is advisable to start with programmers like Python. Python programming language is important because it is suited to machine learning. We will discuss the python program in detail and the other programs that can help you get started in your AI in our subsequent subtopics.

- **Step 2: Learn about machine learning**

  You also need to delve into the world of machine learning. Get to know what it is all about and its importance.

- **Step 3: Take part in contests**

  If possible, you needed to register and take part in any AI or BOT programming workshops or contests within your vicinity. If you cannot find any consider looking for some on the internet.

Reinforcement Learning: Programming Languages for Artificial Intelligence You Need to Know

There are several AI programming languages that are beneficial to you and which you need to know. These includes:

- **Lisp**

  It is a mathematical notation designed for computer programs. It is based on lambda calculus. The program is used to manipulate source code to give rise to rise to macro systems that enable you to come up with new syntax.

- **Smalltalk**

  Small talk is useful in machine learning and in the generation of genetic algorithms. It is also useful in neural networks and simulations.

- **Prolog**

  Prolog programming language is used for symbol reasoning and database. It also a very popular application with AI experts today.

- **Python**

  Python is one of the most popular programming languages which AI researchers use today. Python can be used with packages for applications such as Machine Learning, and Natural Language Processing.

It can also be used with Neural Networks. One key advantage of Python language is that it can be used with various programming paradigms.,

- **C ++**

  C++ is a very popular language program with AI researchers today. The software is a general-purpose programming language created as an extension of the popular C programming language or C with classes. The programming language has grown with time in functionality. Today's C++ complex in nature and is equipped with object-oriented, functional features. It has facilities for low-level memory manipulation. This programming language is often implemented as a compiled language. In addition, the vendors of this program provide C++ compilers such as the Free Software Foundation and IBM.

  C++ was created to favor system programming. C++ has at least one main function. When using the program, you are allowed to divide your code into separate functions but ensure each function is performing a particular task.

  You as a compiler should be able to know the functions name, parameters as well as return type. You can also get various inbuilt functions from the C++ library, which enables your program to call.

- **Java**

  Java is another popular general-purpose programming language with the researchers today. The program is class-based and object-oriented. The program is specifically designed to carry out a few implementation dependencies. With Java, you, as a developer, are enabled to write once and run anywhere-WORA. This will enable you to run your compiled Java application on all platforms that support Java without the need for you to do a recompilation. Some of the latest versions of Java program include the popular Java 13, released in 2019, and Java 11, which hit the market in 2018.

Key Softwares That Are Used for Predictive Purposes

Your career can benefit greatly from the predictive analysis. It can be used to improve overall productivity in your workplace. It is also useful in a reduction of business risks as well as in the detection of possible fraud even before it happens. Predictive analysis is also used to identify and address client expectations in business. This will give you a clear lead ahead of your competitors. Such intelligence is of great help to your business especially when you are generating business strategies. You should, therefore, look for the proper tools to implement them for you to benefit. Here are some of the important predictive analysis software available for you today.

- **SiSense**

  This is a software designed to be used by companies of various sizes. The advantage of this software is that it is user-friendly. You also get various business analytic features that will aid you in staying ahead of the competition. The software has the capability of carrying out data preparations that are complex in nature. They make the data easy for your understanding and enable you to use them to make key business decisions faster.

- **Microsoft R open**

  The software is an open-source platform that specializes in the analysis of statistics and data science. The software was specifically created on the statistical language R-3.5.0. The platform can be used with various packages, applications, or scripts. Some of the popular applications that work with this platform are Windows OS and Linux.

- **Microsoft Azure Machine Learning Studio**

  This is one of the popular platforms used for predictive analytics by AI researchers and other software experts. The platform makes use of data science and as well as cloud-based tools to help you come up with complete reports based on different types of available data. With this platform, you are able to create, use, and share predictive analytics solutions fast and easy. This software is great for startups and small companies.

- **Oracle Crystal Ball**

  The platform makes use of its spreadsheet-based application to make predictive modeling or forecasting. It is also used in simulation as well as in optimization. The platform is best suited for strategic planners or financial analysts. This platform is quite popular with software engineers and other scientists.

Chapter 4: Jobs and Careers After the AI Revolution

The artificial intelligence revolution toughly inclined the biosphere of toil in the 21st epoch. PCs, procedures, and software streamline daily errands, and you cannot imagine how you could accomplish most things devoid of them. Nevertheless, is it also incredible to visualize how you could grip most dealing outpaces without the humanoid workforce? What is by now unblemished and assured is that first-hand technical enlargements will devise an essential impression on the international employment marketplace inside the succeeding limited ages, not just on manufacturing jobs, but on the fundamental of humanoid errands? The following are various application areas after the AI revolution.

Philosophy

Philosophy is a very important field as it attempts to answer important critical thinking questions like can a machine act intelligently? Can it solve like a human being? Is computers' intelligence like human ones? For example, the viewpoint of A, where A is a discipline, comprises theorists scrutinizing the perceptions of A and occasionally commenting on coherent and non-coherent concepts. Artificial brainpower has nearby precise acquaintances with the way of life than other disciplines because it segments numerous conceptions with thinking, like action, awareness, epistemology (what it is serviceable to say about the biosphere), and even open will. The thinking of a most regularly encompasses guidance to the consultants of A nearly what they can and cannot do.

The AI point of assessment is that ethical philosophies are convenient to AI only if they don't inhibit human-level simulated organizations and arrange for a foundation for scheming arrangements with views, go perceptive, and proposal. AI investigation has principally highlighted validating the activities accessible in a state of affairs, and the concerns of enchanting each of numerous activities. To do this, AI has mainly dispensed with pretentious estimates to sensations.

Mathematics

Mathematics is used to write the logic and an algorithm for machine learning. Philosophy thinks and defines a particular intelligence and the way it should work. But here comes the intelligence of Mathematicians to come out with calculations and algorithms for learning. Good knowledge of mathematics is a necessary imperative skill to develop a model of AI. For example, Earl Stanhope's Sense Presenter was a mechanism that was capable of resolving syllogisms, mathematical complications in a rational formula, and straightforward inquiries of likelihood.

In 1815-1864, George Boole brought together his recognized linguistic for creating reasonable insinuation in Boolean algebra. In 1848-1925, Gottlob Frege created a sense that is fundamentally the first-order sense that nowadays forms the utmost basic understanding demonstration scheme. Between 1906 and 1978, Kurt Gödel displayed that there are restrictions on what sense can do. His Incompleteness Statement exposed that in any official reasoning potent abundantly to define the possessions of ordinary figures, there are true testimonials a procedure cannot establish whose truth. And in 1995, Roger Penrose attempts to demonstrate that a humanoid concentration has non-computable competencies.

Computer Science

Computer science is an academic discipline offered in contemporary colleges and universities. A Computer scientist writes the codes for making the neural network for artificial intelligence. It then updates the values or properties of the neural network based on the data provided to the system. You achieve Artificial Intelligence this way. Historically, you base its speculative tradition upon the long-standing behaviors of representative reasoning, calculation, and the moderately more current enlargements in electrical commerce. It was, nevertheless, the speculative toil of the statistician Alan Turing in the 1930s and instigating his philosophy by the statistician John von Neumann in the initial 1950s that tip to the expansion of the modem stored-program PC.

Psychology

Modern Psychology is the discipline that educates how the concentration maneuvers, how do we perform, and how our I.Q.s progress data. It is used to study and find thinking of humans and animals. This discipline enables data science to understand the Brain, Behavior, and Person essential to make things like the human brain. Linguistic is a vital part of humanoid intellect. Much of the primary work on acquaintance illustration links to linguistic and well-versed by inquiries into dialectology. It is normal for us to use the indulgent of how humanoid, and other animals IQs tip to intellectual conduct in the pursuit to build artificial brainpower schemes. Equally, it takes intelligence to discover the possessions of simulated schemes to test the suggestions regarding human schemes. Many sub-fields of AI are instantaneously constructing replicas of how the human scheme activates, and synthetic schemes for resolving real-world glitches, and are permitting valuable thoughts to handover amongst them.

Neuroscience

Tens of billions of neurons make up the brain for each attached to hundreds or thousands of further neurons. For each neuron is a pretentious handling maneuver, e.g., just firing or not firing provisional on the overall sum of action nurturing into it. Nevertheless, outsized links of neurons are authoritative computational maneuvers that can study how top to function. The arena of Connectionism or Neural Links tries to construct synthetic schemes centered on shortened links of streamlined synthetic neurons. The objective is to create dominant AI systems and replicas of numerous human aptitudes. Neural links work at a sub-symbolic level, however much of the cognizant human perceptive seems to function at a figurative level. Synthetic neural links execute well as many modest errands and offer moral replicas of utmost human aptitudes. However, there are numerous errands they are not so virtuous at, and other methods appear more auspicious in those zones.

Ontology

It is the training of the types of possessions that are present. In AI, the curriculums, and stretches contract with countless varieties of substances, and we learn what these categories are and what their elementary belongings are. Prominence on ontology started in the 1990s.

Heuristics

A heuristic is a way of exasperating to determine roughly or a hint entrenched in a package. You use the tenure variously in AI. You can also use heuristic functions in some methodologies to pursuit the quantity of how faraway a protuberance in a pursuit tree appears to be from an objective. Heuristic establishes that equate two nodes in a pursuit tree to see if one is restored than the former, i.e., create up an advance in the direction of the goal and maybe more worthwhile.

Will AI Replace Jobs?

The panic over AI replacing tons of human jobs is misinforming. AI cannot replace jobs. Well, not 100 percent. The work of machines will replace not every specific occupation, but machines will perform some individuals' occupational activities. For instance, the risk of a barkeeper being interchanged is very high. Already today, it is theoretically practical that a machinelike mechanism could blend drinks, direct the clients' orders straight to the kitchenette, collect criticisms, and assent the clients' cash. The mesosphere in the inns or the cafés will no longer be similar. Because of the shortage of approval by probable consumers and the great acquirement costs, it is convinced that a huge percent of all barkeepers will not drop their careers in the succeeding few ages. However, they recognize that no job is safe. The question should not be whether it will change the workplace. The real question should be how companies can successfully use it in ways that help makes humans faster, productive, and more efficient. This will make jobs shift and evolve, not disappear. It will augment how to people complete their work by doing things such as pulling and analyzing data to aid in real-time decision-making. An example of this might be how AI could empower a customer service agent with information from a CRM platform to help the agent decide whether to issue a refund.

A skills gap will emerge among human employees. It will convey many aids in the arrangement of higher efficiency, GDP growth, enhanced commercial concert, and new affluence, but they will also adjust the services essential of human employees. With AI automating repeatable and mundane tasks from accounting to the assembly line, the skill set of the human workforce needs to evolve. The focus and responsibilities of the average employee will widen and deepen. People will need to have a wealth of knowledge and more skills. Essentially, they will need to be great at doing a lot of things.

You need to think about educating our current, and future workforce to learn the skills that AI cannot replace, including:

- **Social skills**: AI will find it is hard to have intercultural sensitivity. They cannot be good at being a leader, cannot take part in a brainstorm or do interpersonal interactions. Cognitive skills: AI is less effective when it has to make judgments based on the specific data on which they have trained it. In the real world, people often decide about situations that do not have previously faced. The problem lies in systems that can match data, but not understand its significance. Skills like complex problem solving, reasoning, negotiating, and decision-making will be important for human workers.

- **Emotional Skills**: AI technology is far from being able to replicate empathy, adaptability, and other emotional skills. Think about this in the form of health care. While AI is empowering doctors and nurses with information to help them decide and complete various tasks, the human touch and conversation with a doctor will always remain important. To come across the perfect principles set for the Commerce forthcoming, personnel must learn new significant recommendations, but they requisite also adapt the educational system to this new context situations. An arrangement at the World Economic Forum states that both schools and universities should not instill the biosphere as it was, but as it will be. They, then, need new stipulation approaches for specific nations. They must inspire scholars' concern in topics like calculation, material knowledge, discipline, and expertise when they are still in school, and tutors with numerical proficiency must impart scholars how to contemplate judgmentally when by means of new broadcasting and help them accomplish a necessary hold of new digital and material strategies.

Creative Jobs

Creative professions have advanced in current eras, and machines will not replace humans in these professions in the upcoming years either. Whether they are celebrities with their tune, performers with their works or writers and thespians with their legendary, or photographic works, or humankind and broadcasting intellectuals, the projection accumulative mandate for their occupations. Mechanism knowledge, as the most powerful branch of artificial intelligence, can only achieve tedious errands by imitating facts and succeeding instructions. It cannot take part in a brainstorm, think creatively, or tackle novel situations, e.g., situations it hasn't encountered before.

Humans can write an emotionally compelling story or soothe a frustrated customer with a personalized conversation, Humans are proactive, and AI is reactive. Humans take a long rational perceptive, passionate aptitude, and a non-linear method to work and life unbearable for AI to attain. That's the bottommost line. The fact relics that AI can't adjust or generate the way people do. AI isn't skillful in identifying novel shapes and new manners or exceptionally grants them. Consequently, there is a level of humanoid originality and complexity required to accomplish and construct upon all software submissions.

In the forthcoming, the end-user will still mandate resourceful showbiz selections and realistically alluring performances. Ever since you encompass no mundane, gifted software can scarcely perform these professions. The alike relate to the systematic expansive sector or occupations with an emotive element. The announcement with other persons will continually come right from individuals. Communication gradually takes place in communal links, but you have to maintain and technically equip them. This is the central challenge of Industry.

Careers in AI and Data Science in The Future

Experts expect Artificial intelligence and data science will be responsible for the most significant and disruptive innovations cutting across the various sectors. AI scientists will play some key roles in this area. One of these roles will be for the scientists to come up with a high-level research-oriented work which surpasses the normally available techniques. Businesses will strive to be part of this action by investing heavily in algorithm-based platforms that assures them an added advantage in the competitive market. This may mean the businesses will apply machine learning at higher levels than we see today.

Moreover, more people will be holders of PhDs and masters' degrees, just like the way we find a big population of MBA graduates today. You should also expect conservative businesses and industries to automate some of their key functions to create a high tech-oriented company culture as well as in their hiring profiles. If the companies fail to embrace this technology, they will miss out on the advantages that come with machine learning. As the designing and improvement of AI machines advance each passing day, experts are already predicting that the economic activities will be disrupted in a big way.

It is also a known fact that AI systems take a big chunk of data as compared to humans. The implication of this is that you might be forced to change your career in the next near future. Although it is expected for the AI to make some jobs and careers better, one cannot ignore the fact that most jobs will be declared absolute by this technology. AI will link the customer to the service or the product, thereby cutting various jobs that exist in between.

What to Study? Artificial Intelligence Operating System

What is an artificial intelligence operating system? well, this is a system that is used to manage the computer software and as well as the hardware. It also helps in providing a common service necessary for the computer or a machine to generate a solution for a complex problem easily. It is expected that artificial intelligence will be playing key functions in operating systems of computers, devices, and other machines. The benefits of AI-based operating systems is quite immense. Key fields such as academic, defense, medical, research, scientific, are expected to benefit from this technological development.

You, therefore, need to be part of the action in order to benefit because the AI learning process is expected to be part of us for the unforeseeable future. Once the development of OS based on AI is complete, it could make your life much more interesting as it will help you perform expert tasks, and monotonous, boring tasks that you normally don't enjoy doing. Another benefit of AIOS is that the computer crashes that you often encounter will be greatly reduced. Besides, the time to execute an operation using AIOS would be much shorter as compared to the time it normally takes in a normal OS.

Moreover, artificial intelligence salaries are incomparable to none. It is one of the fattest paychecks you rarely find even in the most coveted careers. This is because careers in AI are highly competitive and calls for rare specialist talent. To locate such talent is almost an impossible task, that is why possessing such talent will make you a much sought after professional.

Currently, a professional in the field of AI is able to take home a salary ranging from an average of between $100,000-180,000 per annum. This, however, can higher depending on the uniqueness of your qualifications and giftings.

Is Data Science A Good Career?

Yes, data science is a good career. Data science, as the science of extracting insights from data along with domain knowledge, helps in strategic decision-making, leading to better customer experience, reduced costs, and higher profits. With so many data around, governments, corporates, brands, etc. are getting pressure on how to make sense of it and how to use it for more efficient functioning and profits. Data Science is the answer to this challenge.

A data scientist structures big volumes of data then subject it to data analysis. This entails researching both the data and as well as their origin and structure. It is done basically to strengthen the data set, which is incomplete. It is also used to create links between one abstract data set and the other. The work of a big data developer is majorly to deal with the set-up and processing, as well as the storage of vast unstructured data volumes belonging to large companies and governments. On the other hand, data artists are charged with the responsibility of graphic presentation and editing of data volumes.

Although data scientist career is relatively new, its importance cannot be gainsaid. It is one of the very important careers in the field of communication and science today. It is widely expected that this new career will be one of the hottest and well remunerated in the near future. If you are considering pursuing a career in this field, then it is advisable that you acquire key IT skills that will help you maneuver through the complex demands of the career. You need to be adept in the relevant programming languages. You should also have the necessary knowledge and skills to write complex programming codes. It is also helpful if you familiarize yourself with the business processes of the corporate world in order for you to create reasonable links. Furthermore, you need to acquire basic knowledge in business fields such as business administration, economics, and marketing. Besides, it will help your career if you cultivate rich interpersonal qualities.

A data scientist is expected to offer quality services, which meet the expectations of their customers as well as those of the employer. For you to be successful in this field, you must receive the requisite training that makes you highly competent and effective. During the course of your work as a data scientist, you are expected to perform data processing as well as the computation on big scales. For you to be adept in this, consider getting training from the relevant institutions. Such training entails some knowledge in engineering, mathematical sciences, and social sciences.

Computing and Artificial Intelligence

Computing systems make use of machine learning and natural language processing. It also makes use of AI data mining methods. However, computing systems doesn't stop there; it strives to imitate the key functionaries of the human brain on a machine. It makes use of available data to make key decisions, just like the way human beings make logical decisions. Once it is through with its analysis, computing systems, for example, IBM then gives its best alternative to a given problem at hand. This might not be the right choice; it is, therefore, upon you to decide the appropriate course of action is in a given situation.

The key difference between computing platforms and artificial intelligence systems is that AI is created to carry out some tasks on your behalf while the computing system serves the purpose of giving you the advice you need or the guidance you need before making key decisions.

Businesses can use the platform to come up with risk factors needed before a decision is made. The platform provides companies with a recommendation on investments as well as other key business decisions. The opportunities offered by this technology is huge and is largely untapped.

Chapter 5: Artificial Intelligence Algorithms

Over the past years, algorithms were believed to be a subject for mathematicians and computer scientists. However, with the recent technological advancements and the rapid evolution of artificial intelligence, it has become necessary for everyone to have a glimpse of what algorithms entail. AI has been in the present-day adopted in almost all areas of operations, including hotels and hospitals, where it is used to make human activities easy by using machines to handle complex tasks.

What then does the term algorithm in AI mean? An algorithm is a basic set of rules that are followed mainly by computers in carrying out calculations or handling other problem-solving operations. The goal of an algorithm is basically to solve a problem using the simplest way possible and within a short time by following a predetermined procedure or set of steps. You can also view algorithms as shortcuts that help you give instructions to your computer. By adopting the use of these algorithms, you communicate to a computer by telling it what action to take next using statements which are denoted by "and," "not" and "or." These statements guide the computer in handling problems, just like a human being would have done. Most of the algorithms resemble mathematical problems, which begin as easy tasks with increasing levels of complexity as they are expounded. The simple steps you follow in your house when baking your favorite cake depicts a simple form of an algorithm.

With the growth of the AI sector in trying to bridge the gap that exists between human and machine capabilities, it is essential to understand the various algorithms used in AI. It is, however, necessary to note that not all algorithms are applied in artificial intelligence. The commonly used algorithms here are:

- Convolutional Neural Networks (CNNs)

- Recurrent Neural Networks (RNNs)

- Reinforcement Learning (RL)

You may be wondering what is meant by the term neural networks in algorithms. This should not send chills down your spine because this chapter contains a detailed outline of everything you may want to know concerning algorithms and their applications in AI. Neural networks are a unique set of well-defined algorithms that are tailored to resemble a human brain with the ability to recognize patterns efficiently. The designs they recognize are usually in numerical form and contained in vectors in which all data such as sound, image, and text must be translated. These neural networks interpret sensory data through perception, labeling, as well as clustering of raw input. Different types of neural networks have various functions, such as clustering and classification of data based on the similarities depicted.

Before getting into the types of algorithms, you need first to understand the essential layers found in a simple neural network.

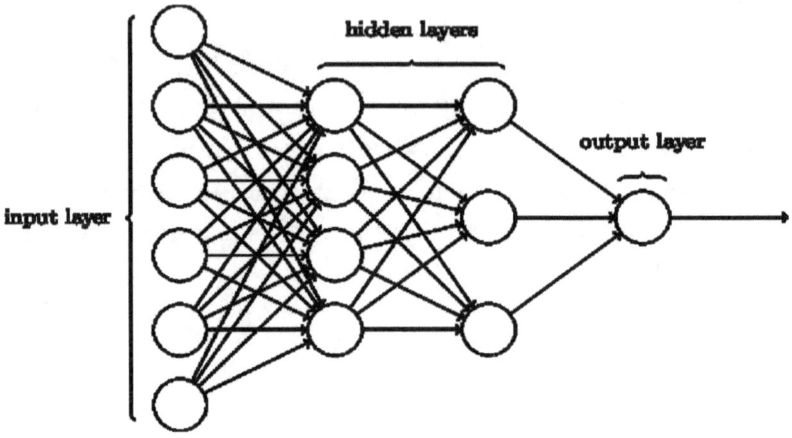

Figure 1: The architecture of a simple neural network

Input layer - here, you input data for your model. The amount of neurons introduced here is similar to the number of features present in the data in the output layer. For instance, the sum of pixels in case you are using an image.

Hidden layer -it is a layer that comes directly after the input layer. Based on the algorithm model as well as the data size, the number of hidden layers may be more than one with each one of them containing a different amount of neurons that are more than the sum of features in the output layer.

Output layer - contains output generated from all the hidden layers. The output information is manipulated using sigmoid logistic as well as softmax functions that convert each output class into a probability score of the representative type.

Convolution Neural Networks (CNNs or ConvNet)

CNN is an algorithm that can take the input, for example, an image, allocate weights to its various features, and still be in a position to differentiate the aspects from each other. The pre-processing elements required in CNNs tends to be lower in comparison to those required for other forms of algorithms. Additionally, with sufficient training, CNNs can learn the various types of filters rather than having to hand-engineer them as it is the case with other types of algorithms.

The design of CNN is made to resemble the connectivity of brain neurons in a human being. This architectural design is based on the organization of the human visual cortex. Within your visual cortex, different neurons react in response to varying stimuli within a restricted section of your visual field, referred to as the receptive field.

Layers Found in ConvNets

ConvNets are made of sequential layers, with each layer transforming the input data from one volume to another by the use of differentiable functions. The primary role played by the ConvNets is to minimize the size of an image into a smaller form that is easy to process while still maintaining its critical features. Below are the common layers found in CNNs.

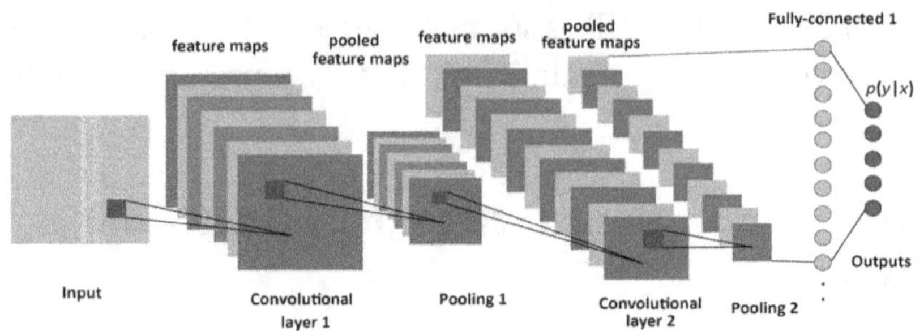

Figure 2: A sample CNNs architecture set up

Input layer - this contains the raw data input, that in most cases, is usually an image. Let's assume that your input is an image of volume 32*32*3

Convolution layer - this forms the core building block of a CNN where all the complex computations are done. Here, the output volume is computed between filters as well as image patches. The amount of filters used determines the depth of the resultant output volume. Suppose you use a total of 12 filters, then the size of your output will be 32*32*12.

Pool layer - This layer is added in the model with the sole purpose of lowering the overall volume of the data under computation. There are basically two types of pooling that are; maximum pooling as well as average pooling with maximum pooling being commonly used in CNNs. Maximum pooling suggests that the maximum element be pooled. For example, in your illustration, if you use a maximum pool containing filters of 2*2 and a stride of 2, then the resulting output volume will change to16*16*12. There may be several pooling layers based on the size of the entire computation.

Fully connected layer - contains the input generated from all other segments, and it computes output volumes of class scores in sizes that are equal to the respective sum of classes.

CNN's are applied in various areas of AI, including image processing and others such as:

- **Text recognition** - CNN has been adopted in decoding visual images in texts through a function known as optical character recognition (OCR).

- **Facial recognition** - apart from image recognition and processing, CNN has also been used in the field of facial recognition. Recently, it has been seen to handle the task of recognizing faces from different angles, even when the visibility is limited.

Recurrent Neural Networks (RNNs)

Anytime you are thinking. You do not start the process from scratch. Whatever you think, comes from some information that already exists in your mind. Similarly, as you continue reading through this chapter, you will understand some words, based on your previous encounter with them. This simply means that you do not discard all the information you have in mind once you are done using it, but instead, you store it for future use. This simply means that output from one process may be used as input for another process. This is the concept behind which RNNs operates.

RNNs is an artificial intelligence algorithm that uses the output information from a previous layer as input in the current layer. Here, the input, as well as the output vectors from the different steps, are dependent. Just as the name recurrent suggests, the dependent input and output keep on being used over and over again. That is, they recur. This type of algorithm is primarily used in language processing and speech recognition, where it recognizes the sequential characteristics of data and uses specific patterns to predetermine the next probable action. The traditional forms of algorithm considered the inputs and the outputs to be independent variables. This, however, posed a challenge, especially where a prediction of the next action or step is required. RNNs were developed to bridge this gap through its unique feature, which is the hidden layer, which acts as a memory to remember certain bits of information concerning a sequence. RNNs can accommodate more than one input vector and work on them to produce one or more outputs based on the weights as well as biases applied.

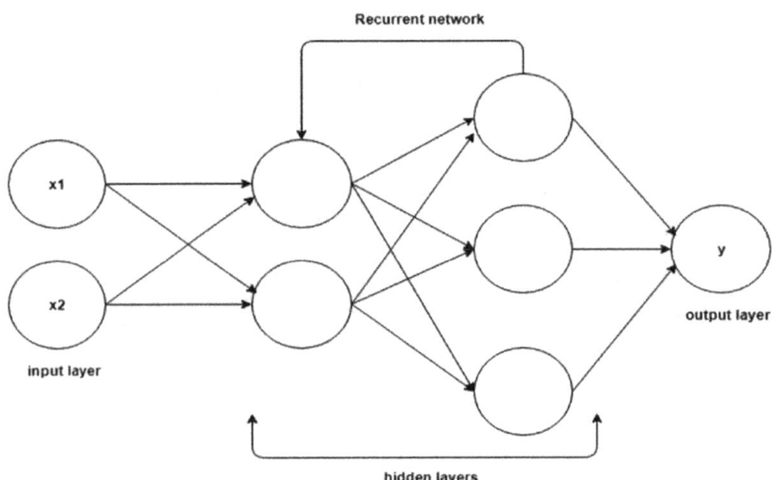

Figure 3: A typical architecture of RNNs

The hidden layer remembers all the previously handled information. RNNs adopts the use of a similar parameter for every input while still performing a similar task for the rest of the input sequences to bring out the output or outputs. The chain-like nature of the RNNs shows that there exists a relationship between the sequences and lists of the various inputs. The RNNs poses a feature which makes them more preferred for use than the CNNs, that is they allow you to operate over a series of vectors in the input or output layer or in general cases both layers unlike the CNNs where they only accept a fixed size of vector input and produce a fixed volume of vector output.

How Does RNNs Work?

To help you understand how the RNNs works, here is an illustration.

Assume that you are working on an input network consisting of one layer of input, two hidden layers, as well as a single output layer. Every hidden layer has specific biases as well as weights applied to them, for example, hidden layer 1has weight 1 (w1) and Bias 1 (b1) while hidden layer 2 has (w2 and b2). This shows that the two hidden layers operate under different weights as well as biases and are thus independent of each other. This independent character means that the second hidden layer does not rely on the output from the first hidden layer. Such a scenario, therefore, does not qualify under the RNN.

When the RNN model is introduced, the following happens:

The independent activations will be converted to dependent activations by adding similar weights as well as biases in the hidden layers. This means that the output from the first layer will be used as the input in the subsequent hidden layer. By converting the activations from independent to dependent, the two hidden layers can then be easily joined together by the similarity of the bias and weights in them to form a single recurrent layer. The recurrent network helps in memorizing the output from the previous steps. You should always remember that the recurrent network is the unique feature that distinguishes RNNs from other forms of algorithms.

RNNs in AI has been adopted for use in a variety of fields such as:

- **Machine translations** - RNN is applied here to help translate texts or statements from one language to another without altering the original meaning.

- **Speech recognition** - though the input of specific sound waves, RNN is used to predict the phonetic segments which assist in generating words.

- **Generating texts and language modeling** - this follows a sequence of texts introduced in the input layer to predict the next likely word.

- **Image descriptions** - RNN works well here when combined with CNN. The CNN does the segmentation of the image, which is fed into the model at the input layer while the RNN uses the so segmented data to recreate the image descriptions.

Based on the outline concerning RNNs, you can undoubtedly agree that it is designed to function and operate in a way similar to the human brain. This feature makes it easy to apply it in AI for purposes of data and image processing.

Reinforcement Learning (RL)

The main idea behind the reinforcement learning algorithm is learning from interactions with the environment. For instance, assume that during winter, the weather outside is chilly, and your body is freezing, you have to look for some warmth. You then decide to light up fire at the center of your house to keep you warm. Your little sibling then joins you to get some heat too. He approaches the fireplace and feels warm. This makes him understand that fire is a positive thing. The warmth from the fireplace draws him nearer to it, and he eventually decides to touch it but ends up being burned. Ouch! He understands that the fire is not too friendly as he thought. It comes to his understanding that the fire is helpful when you are at a safe distance from it because it gives you warmth, but moving too close to it may get you burned. The same concept of learning applies to RL.

In AI, RL is a type of programming used in training algorithms based on a system of reward and punishment. The software agent here learns through having interactions with the general environment in which he receives awards by taking corrective actions and gets punishments for incorrect responses. In our illustration, the reward is heat or warmth gained by keeping a safe distance from the fireplace while the punishment is getting burned by getting too close to the fire.

Basically, in a simple RL set up, the agent observes his environment, then decides on the best course of action to use in interacting with the existing environment. The result of the actions taken may give the software agent a reward or a punishment. Therefore, in every action taken by the agent, he aims at maximizing his reward.

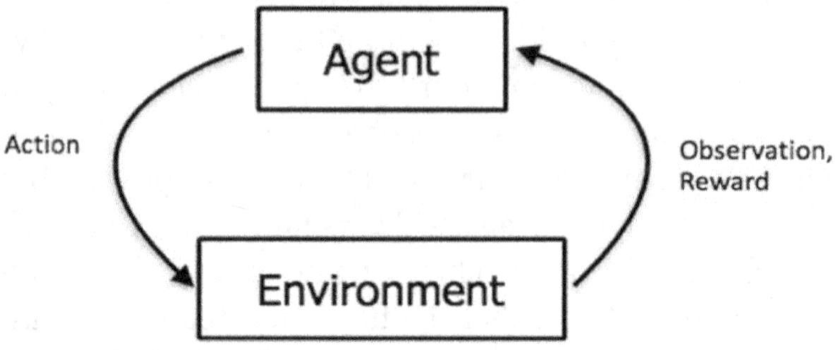

Figure 4: A simple RL setup

The unique feature of RL that makes it stand out among the other types of algorithms is the training of the agents. Here, rather than using the programming data provided, the agent is required to interact with the general environment on its own in an attempt to help it learn ways to maximize rewards while minimizing punishments.

RL has gained popularity in the AI world today following its applications in various fields such as:

- **Medicine** - RL has been applied in the field of medicine to perform clinical trials as well as drug therapies.

- **Gaming** - following the victory of Alpha Go, a machine learning algorithm over a human player in a game, many people have adopted the use of RL in games of all kinds. RL is the algorithms that are mostly used in computer games.

- **Robotics** - owing to the fact that RL can take place without any supervision, it becomes the best option for use in robotics. Its application in the robotics industry has brought with it an exponential growth in robotics.

- **Autonomous vehicles** - RL has been adopted for use in autonomous drones, vehicles, ships, and trucks. The use of reinforcement learning has resulted in considerable growth in the autonomous vehicle industry.

As the position of RL in the AI industry continues to grow, companies will be required to invest more in data and resources to help them understand the various ways in which they can implement this technological advancement in their products, services as well as industrial operations.

Artificial Intelligence and Image Processing

You might be wondering what image processing entails. Image processing refers to the manipulation of an image to extract some information from it or to enhance it. Most images that are taken using regular cameras may be misfocused or even contain lots of noise, and this calls for processing. Image processing may come in the form of edge detection or filtering. From the detailed explanation of AI algorithms, it is evident that there exists a strong link between AI and image processing. Most of the AI algorithms, such as CNNs, RNNs, and RL, are commonly adopted in image processing.

Chapter 6: All in Real Life

With the coming up of Artificial Intelligence, a lot of industries can enhance CX by studying more about customers and forestalling their needs. A lot of CX focused trademarks are arraying Artificial Intelligence technologies tactically at main consumer touchpoints. There are examples of AI-powered CX from various companies that will show customer experience can be an algorithm for accelerative thinking industries.

- **When it comes to retailing, AI personalization unlocks access to 1% customer**

  The information indicates that the top 1% of retailer's buyers are valued 18x further than the average buyer. The best method you can enact for those discerning, high-value buyers is by personalization. Usual personalization such as user-specific page outline is table stake for memorable buyer experience, and thrilling customization is required. This where you will need an advanced machine. Thrilling personalization goes beyond one-off personalized newsletter to a buyer tailored promotions that are distributed at the correct time, to the right device, and with an impeccable message. You can see it as a move from buyer segments to the audience of one.

- **The building of trust and loyalty in a global bank**

The Royal Bank of Scotland is handling 17 million customers transversely seven trademarks and eight, unlike channels. When you look at its history, their strategy concentrated on aggressive sales goals intending to upsell buyers into current credit cards. From the buyer look of things, this has brought about a heap of digital and paper spam. Royal Bank of Scotland has decided to revamp its partnership with the buyer turning to AI to improve customer experience. The approach was taken to lever data wisdom into entirely new levels of purchaser contact. When a customer severally overdrafts an account, AI flags applicable bank staffs to contact the customer with financial guidance.

- **The specific airline, new data intelligence is driving CX innovation**

Air Canada is handling 45 million customers yearly, having many people booking online or using a mobile application. They are seeking to improve understand their customers and also enhance the mobile app experience. They have enacted an AI and machine studying info analytics system that gives insight to their customer characteristics through digital and offline frequencies. Company heads leveraged the info analytics vision into customer-facing enactment improvements and streamlined website familiarity.

- **AI battles ticket bots in the entertainment sector**

The Internet has changed the dynamics of live events by coming up with easily accessible secondary ticket marketplaces like Craigslist and StubHub. Recently automatic bots purchase several tickets, depleting supply, and instantly offer the tickets for sale at main markups. Ticketmaster has changed AI to rewrite rules by the use of a machine learning system known as Verified Fan. This system encourages those buying tickets first to register their interests before tickets go on sale. AI systems analyze each registrant to identify scalper bots behind the scenes. This has resulted in 5% of tickets sold using Verified Fan getting to the secondary marketplaces. Most of the people, more so artists, and fans are so happy with the way of acquiring tickets.

- **Some hotel brand, new insight requires new AI**

To understand customers well, the hospitality industry has enhanced techniques such as mystery shoppers and customer assessments. Leveraging content of many online assessment sites has been seen as difficult or expensive. This was there until the hotel brand Dorchester Collection created a custom AI analytic system that is an essential giant focus group operating endlessly in real-time. The system has been capable of bringing along 75000 guest reviews from 28 hotels transversely 10 trademarks and send its outcomes with a 30-minute film.

AI Changes in Market Place

- **Buying and selling faster regards to AI**

  There is a belief that AI can change the way humans shop and are full of joy because of the potential opportunities and worth it's bringing to those buying and selling. AI is improving the marketplace, thus making it more efficient for buyers and giving a more natural interaction between sellers and buyers. AI has automatically enhanced the quality of pictures and translated listings and messenger interactions. There is an introduction of new features that use AI for price range suggestions and automatically categorize. This means that when you want to sell a good, a marketplace can use AI to assist you in selling it quickly by giving options about your price based on comparable amounts of the good. It will also categorize the proper based on the picture and how it is described, making it easier for you. There is an ongoing test on camera features that can use AI to suggest goods you may interest in having, i.e., when you like your friend's shoes, you can have a picture, and marketplace's AI tech can recommend the same product on sale around you.

- **A lot of shopping options within you**

  Adding to the new AI features, there has been an addition of a variety of contents from trades inclusion of vehicles, house rentals, home operations, and shopping from e-commerce dealers. Nowadays, cars are the most vital category for the marketplace worldwide, along with furniture and electronics. After coming up with vehicles on the list of local dealers a year ago, the market has been considered one of the leading destinations for people to acquire and sell already used cars.

- **Building a safe and trusted community**

Shopping online needs a lot of confidence in the people and trades you are buying from or operating with. This is why you should enhance features that enact a safe and trust community, including:

- Detect and delete objectionable content: due to AI technology, you can detect and remove items that are violating your policies by analyzing pictures, content, and context.

- Rating buyers and sellers: buyers and sellers can rate each other to know if they have a good or bad experience and leave a comment on what can be improved and its reliability. It also helps people be informed on who to contact, community ratings that assist in the creation of great experiences by enacting good character.

- A lot of Robust reporting tools: the community has helped to make buying and selling better for anyone by giving content that doesn't go along with the marketplace. You can report the listing of items that are violating commerce policies. You can also inform buyers and sellers doing illegal activities.

Just Getting Started

When you look back the past two years, you will be inspired by the people across the world using the marketplace to do great activities. There are stories like that of Rudolph and his family who bonded with his family and community after the loss of his wife through cancer by selling birdhouses for a cause. They have continued to work on delivering new useful features that you have an interest in.

AI Changes in Time Management

Most of us are not capable of working on everything we planned to execute in the morning. It's on rare occasions when someone can perform all the plans on the list. This will pile up the tasks daily and weekly. This is a point where time management will be changed to time intelligence when connected to artificial intelligence. You can employ AI to assist in the cumbersome lifting process of putting together data and context. This will give it a centralized system.

Automated Information Harvesting

AI technology permits automated info harvesting to have a business plan. There is control of manual time inputs from the workers; thus, it automatically collects data from the trade ecosystem, holding it back to particular deliverables and the limitations. It provides you with facial identification technology where a worker can walk to their time clock and log in by the use of facial identification. Due to this, the process tends to be faster, accessible, and gives advantages of biometrics without any added course or software costs.

Receiving Assistance from Personal Assistants

This is one of the most familiar kinds of AI that you have come across. Alexa, Google Now, and Siri are usual AI-powered systems that can study your demands and characteristics. They will provide you with essential data when asked a question. Other AI manufacturers keep their focus on creating a wise to-do-list that can assist you in prioritizing your tasks and sending notices. They can give a record of how they have been performing previously. Others can also help you plan your emails, alert you about traveling, manage your contact lists and pictures.

Manage Task in Real-Time

With info becoming real-time, trades have sorted to make crucial decisions immediately. Heads-Suite and managers can have advance notices and alerts that cause them to have a pro-active approach while making verdicts.

Customizing Workflow

A lot of trades are having problems when it comes to timesheets, the trouble of exceptions, mistakes, and unpredictability. Wise workflows simplify the trade processes by validating the real-time data with built-in exemption; thus, an admin can see where and when it brings a positive impact.

Artificial Intelligence-powered time management tools work in different ways, like:

- **Habit** – AI-enabled tool will note down your habits and give suggestions on due dates, preferably. In case you reply to your mails on Monday morning, then the due response can be set for that particular day.

- **Task priority** – the tool can estimate the urgency of any kind of task by tracking the information unknowingly from the to-do list.

- **Keeping a record of weekdays and weekends** – the tool can store files of duties that you can execute on weekdays. It will also schedule tasks to be done weekly.

- **Upcoming tasks** – if you are full of tasks on a Thursday and you are free on Friday, then AI will try to balance the load throughout the seven days appropriately.

- **Daily and weekly visions** – you can plan your daily and weekly perceptions depending on the tasks you want to execute daily and weekly. The time management tool can assist the due dates, thus helping you meet explicit goals.

You find yourself able to define business goals and create a data model to enhance and give feedback; then, you can use AI for resource optimization and preparation. This can help transport and have logistics in your trade. Forecasting is the best method you can employ to help exceptional tuning project execution and reduce the number of misses. When you decide to use Artificial Intelligence in improving your project enactment, you need to get ways to try and enhance. Most of the AI-powered tools are capable of learning from habits by the use of artificial intelligence, suggesting any necessary changes.

AI Handling Customer Retention and Interaction

Most of the successful marketing sections should ask themselves questions about what their customers want and what kind of goods will suit and meet the wants of a customer. They can answer such questions by the use of artificial intelligence.

Improvement of Human-Computer Interactions

Natural language processing and machine learning are capable projects around the vast area of AI that can be used to enhance and improve the product to be discovered and overall buying experience. Speech identification, along with national language understanding, among others, is capable of options that will allow people to use spoken or scripted sentences to communicate with computer systems. This will enable customers to interact requests directly, unlike going through other interfaces on a device like a mobile phone, table, among others. NLP changes the spoken word to have the intent of the user and when used in partnership with the user's context, can make it to like its more than artificial. Cognitive methods, like deep learning, can increase the process. Advance supervised and unsupervised ML methods can be put in use to enhance advanced models and algorithms. Due to this, commerce systems improved by ML, and AI will wisely cross-sell and up-sell compelling proposals. This will bring about an increase in conversation rates, general order worth, and buyer loyalty and buyer lifetime value.

Assists in Identifying the Needs of a Customer

A lot of many marketers are taking advantage of massive data and predictive analysis for presentation. The logic that the larger the information sets and there is an increase of powerful algorithms being used thus high commendations from customers. Personalized engines will put together personal character with microdata tracking their characteristics along with touchpoint trademarks to make a decision about products and services based on previous needs. This gathers customer data with internet information basing on ML analysis that will incorporate things such as time spent when requesting, clicks, scrolls, and interest levels. When many people are using the Internet and purchasing utilizing the trademark, then there will be better and easy personalization. Buyers will be free to provide insights to personal interests and considering the purchase of the systems.

Improves Your Pricing and Makes It Dynamic

The biggest problem faced by pricing managers is the daily fluctuation. E-commerce companies have platforms that assist in the management of supply and demand with ML that can partition customer buyer segments. This will provide the trade with the power to pinpoint the cost of customers who can pay at a personal level. It also examines customer data and confirms the price a customer is able to pay. Overlooked benefits of AI changes the story in e-commerce and service operation. AI unlocks the hidden pattern in programmer characteristics to learn or study the steps to make or the best opportunities for success. When you are leveraging NLP, the buyer will also get experience, and trade gains profitable insight. This will bring a win for the customer and the brand.

AI Application in Common Life

- **Medical and Chemical Research**
  - **Medical imaging**- medical scans are systematically gathered and kept for some time until they are available to train AI systems. AI cuts the cost and time in scans, thus giving the advantage to have better treatment. AI is providing results that are encouraging to detect pneumonia, cancer, and eye diseases.
  - **Surgery** – robots that are controlled by AI are being used to handle specific tasks during surgery, such as tying knots to close wounds.

- **Sex Industry**

  Testing a dataset that is the same as the training dataset containing silhouettes shaping people to copy real images in human trafficking cases that victims use a black overlay. Systems have been trained to identify hotel-based rooms décor. Most of the top pictures retracted are from the right hotel chain.

- **Transportation Industry**

  It will ensure safety for all road users: the safety of passengers, pedestrians, and drivers is one of the biggest concerns in the transportation industry. AI has decreased the number of human errors. It has been able to predict and monitor traffic. Prediction ways will assist you in performing traffic conditions and automatically calculating alternative methods. There will be vehicle maintenance that predicts using intelligent sensors and dashboards. There will be an implementation of driver characteristics that check and enhances safety and increase production. It can analyze sensor information and check your fleet through data-driven dashboards for wise trains, ships, and trucks connectivity. AI can adjust your route to avoid congestion and incidents and allocate time for occasions where jams are unavoidable.

- **Teaching Industry**

AI has brought about a high impact in the teaching industry by simplifying administrative duties. It has an automated expedition of administrative tasks for teachers and institutions. AI has come up with great ways of grading systems and has gained a lot in the school admissions board. Smart content will be there to give students techniques in achieving academic success. AI technology has technology that has been utilized in students, and they are being taught in schools. AI tutoring systems can give natural feedback and work with students directly. The methods are inception and can become digital teachers and assist students with educational wants. AI enacts global learning because education has no limits, thus AI eliminating boundaries. AI has enhanced IT processes unleashing new efficiencies.

- **Agriculture Industry**

Agriculture and farming is a very vital important profession worldwide. AI has helped in analyzing farm information. Farms give lots of thousands of details on the ground every day. AI has assisted farmers in analyzing a range of things in concurrent like meteorological conditions, high-temperature usage of water from their farmstead to augment enhanced pronouncements. AI expertise assists agriculturalists to have plans to create a lot of yields by determining the type of choices, hybrid seed options, and resourceful utilization. Agriculturalists are exhausting AI to mend recurrent models, thus improving precision in agriculture and increase production. The models can predict weather patterns months and help farmers to make decisions. Seasonal forecasting is valuable for small scale farmers as it helps in improving the countries as their information can be restricted.

Possession of the small ranches to carry on working and developing yields are vital because they help in the production of 70% of the world's crops. AI tackles labor challenges, and this is because of the workforce challenge. Most farms need many laborers to help harvest crops and keep the farms productive. A solution for assisting the shortage of workers AI agriculture bots. The bots have been used in augment of human labor using different forms. The bots can do harvesting at higher volume and quicker speed than social workers. The bots do work accurately by identifying and eliminating weeds. They have reduced the costs of farms by working all through, unlike the human workforce.

AI Changes in Bank System for a Customer

Banks have a hard time trying to reduce costs, meet margins, and exceed customer expectations through individual experience. To enhance this, the implementation of AI is very vital. Most banks have begun embracing AI and connected technologies worldwide. A study by the National Business Research Institute, states that 32% of financial organizations are using AI by the use of voice recognition and predicting analysis. Coming up with mobile technology has a large playing field in the banking sector. Automated AI-powered customer service has gained a firm grip. AI features have enabled services, offers, and insights within programmers' characteristics and necessities. Cognitive machines have been trained to advise and interact by giving an analysis of programmers' data. A lot of banks are deploying tools to help in examining transactions in real-time.

Examples of Artificial Intelligence

Email Spam Filters

The filters are using machine learning, forming AI to learn which emails you want and don't. That is not the only AI you can have in your inbox. Google has come up with an AI-powered service known as Smart Reply that has created a short email message that has helped to reply suggestions basing on the respondent to familiar messages.

Collaboration Tool Slack

Slack is a chat tool that is being used by trade organizations for putting together communication. A lot of organizations have never realized that slack is using AI behind the scenes to help in analyzing of information it has gathered about every company and its workers using the tool. This is, therefore, enhancing developments and improving the production of people using it.

Teslas Studying from Other Teslas

Tesla cars have connected and learned from one another, even though Elon Musk has ruffled feathers in the AI world. In any case, a vehicle moves to avoid hurdles on the highway; the other fleet will get to know what to do after the updates are delivered, sharing information. The team is still working to build full automatically cars for your road, and AI is part of their operation.

Video Games

Artificial intelligence has been associated with a video game for a very long time, and now more of its applications are sophisticated. When it comes to game series, AI individuals have revolved basing on their communication with players. Nowadays, all the video games played are empowered by AI in some way.

Music Compositions

When you want help to focus, relax, meditate, or sleep, you might try out Brain.fm, this is the most advanced AI music writer in the world. Most of the music in the service is created by AI and has been tested for its results to create music. The brain craves to achieve the desired outcome, and this is not the only way AI is making steps in the airwaves.

Self-Driving Cars

One of the significant applications AI innovated autonomous vehicles. The idea was earlier on a sci-fi fantasy, and it's now a practical reality. Most people were skeptical about the technology when it was starting, driverless cars have already been introduced in the transportation industry. Automated taxis have already begun to be operational in Tokyo. For safety measures, a driver is supposed to sit in the car to control the vehicle in case of an emergency. The founder of the cars has said that the technology will help in reducing the cost of taxi operations thus will help increase public transportation modes in remote regions.

American logistics have embraced automatic trucks to harvest a lot of benefits. With the coming of autonomous trucks, maintenance and administration values will reduce by 45%. The majority of companies are still working on their pilot projects, thus striving to enhance self-driving vehicles flawless and safe for passengers. This technology has evolved, and self-driven cars have already gained high confidence and have been mainstream in the consumer realm. The vehicle has a combination of different sensors to perceive its surroundings; such sensors include radar, sonar, and inertial measurement units.

Chapter 7: Data Science

Data science is an interdisciplinary mix of data logic, algorithm development, and technology to unravel scientifically difficult issues. At its core is a collection of raw data, streaming in and stored in company data warehouses. Data science is a discipline that unifies statistics, information analysis, machine learning, and their common methods to understand and analyze real occurrences with data. Data science employs techniques and approaches drawn from several fields in the context of arithmetic, statistics, engineering science, and data science. Data science is alluded to be a "fourth paradigm" of science (empirical, theoretical, process, and currently information-driven). The entirety of science is dynamic due to the impact of information technology, therefore, causing the data flood.

Data science is about using this discovered data in creative ways to get value for your business. We can build advanced capabilities with it. We can learn and build advanced potentiality from the data that we mine.

Data science is an umbrella term that encompasses information analytics, data mining, machine learning, and several other connected disciplines. Whereas a data scientist is anticipated to forecast the future supported by past patterns, information analysts extract meaty insights from varied information sources. A data scientist creates queries, whereas a data analyst finds answers to the present set of queries.

Data Mining, Analysis, and Insight

Data analysis and data mining belong to the business intelligence (BI) subset that additionally integrates data warehousing, online analytical processing (OLAP), and database management systems. Data analytics likewise encompasses a couple of completely different branches of broader statistics and analysis that facilitates combining various sources of data and find connections, therefore, simplifying the results.

These facets of data science are each regarding discovering findings from data. Plunging in at a crude level to mine, observe and understand complicated trends, behaviors, and assumptions. It is about regression analysis of buried insight, which will facilitate corporations to form smarter business decisions.

You have four V's that are used regarding big data; volume, veracity, velocity, and variety. Another V exists that is vital for your knowledge- validity. There is a proverb in programming– "Garbage in, garbage out." the standard of your analysis depends on the standard of your data.

The collection of data might constitute conducting surveys, polls, or doing different experiments. Throughout this collection, data might increasingly become contaminated, which can yield incorrect analysis or contribute to creating the wrong business decisions. While ways of research could contradict by subject area, the optimum stage for deciding applicable scientific procedures happens at the beginning of the analysis method and should not be a second thought.

The increasing data being generated annually maintain obtaining useful information that is a lot more vital. The data often is kept in a data warehouse, summarized data from internal systems, and data from external sources, a repository of information gathered from varied sources, together with company databases. Analysis of the report includes simple query databases and recording, applied math analysis, data mining, and a lot of complex inter-dimensional analysis.

As an example:

- Netflix mines data on movie watching patterns to know what drives client interest and applies that data to form selections on that Netflix original series to introduce.
- Target identifies the customers' distinctive shopping behaviors among their major customer segments that help to guide electronic communication to an entirely diverse market.
- Proctor & Gamble utilizes statistical designs to clearly perceive prospective demand, which facilitates set up for production levels more successfully.

How do data scientists excavate insights? They begin with data research. When presented with a difficult query, data scientists become investigators. They investigate the evidence and attempt to decipher patterns or characteristics among the data. This skill needs some analytical ability. Then, information scientists apply the quantitative technique to understand the information at a more profound level- for instance, segmentation analysis, presumed models, statistical forecasting, artificial control experiments, etc. The intent is to scientifically put together a forensic overview of exactly what the data is admittedly putting across. Data-driven insight is pivotal to providing clever, calculated steerage. Therefore, data scientists assume the role of consultants, offering insight to business stakeholders on how to work on the findings.

The purpose of collecting company data jointly in one structure- generally in a company's data warehouse- is to expedite analysis, so data that is collected from a range of various company activities are applied to enhance the understanding of fundamental trends in their business. One of the fastest-growing areas in data science is most frequently related to multidimensional analysis.

An effective data repositioning strategy needs a robust, nimble, and straightforward way to cultivate advantageous information from collected data. Data mining tools and analysis use quantitative methods, pattern recognition, correlation finding, cluster analysis, and associations to research data with minimal or no IT interference. The ensuing information is then given to the customer in a comprehensive form. The processes are jointly referred to as business intelligence. Managers will choose from many kinds of analysis tools, together with reports and queries, managed query environments, and OLAP and its alternatives. These are supported by data mining that develops patterns that will be applied for subsequent analysis completing the BI method.

Data Types

Data types are a crucial statistics concept that must be understood to exploit statistical analysis to your data correctly and to draw accurate inferences regarding the data adequately. The first factor to try to do after you begin learning statistics is becoming aware of the data variables that are used, like numerical and categorical variables. Differing variables need differing kinds of applied math and visualization approaches.

Having a decent knowledge of the various data variables, also known as measurement scales, could be a fundamental requirement for performing Exploratory Data Analysis (EDA) because you can only apply specific statistical measurements for particular data types. You further ought to understand the data variable that you will apply to settle on the correct visualization technique. Data types should be considered as the simplest way of classifying different variables. Two main types of variables or measurement scales are:

Numerical

Numerical data is measurable information, and it is, of course, data described as numbers and not words or text. Like a person's weight, height, blood pressure, or IQ. Or perhaps the data is a count, like a number of stock shares owned by an individual, the number of chapters in a literary piece, or how many books you will be able to read of your favorite author before the end of the year. (Statistic experts conjointly refer to numerical data as quantitative data.)

Nominal data are often additionally split into two types: distinct and continuous. Note that nominal data that has no order; so, if you would alter the arrangement of its values, then the effect would remain unaltered.

Discrete Data

Discrete data has a logical conclusion to it. It represents things that may be counted; it makes out probable values that may be recorded. The list of actual values could also be finite (also referred to as fixed), or it is going to start from 0, 1, 2, on to eternity (making it computationally infinite). As an illustration, the quantity of tails in one hundred coin tosses assumes values from zero through the finite case to one hundred. However, the number of tosses required to arrive at one hundred tails assumes values from one hundred onwards to eternity (if you never arrive at the one-hundredth tails). The possible values of the coin toss are then listed as one hundred, 101, 102, 103, and so on (representing the countably infinite case).

Continuous data

Continuous data does not have a logical end to it. It represents measurements; its possible values cannot be counted and might solely be delineated with intervals on the real number line. To illustrate, the precise quantity of liquid petroleum bought at the pump for cars with 30-liter tanks would be continuous data from zero liters to thirty liters, described by the 0 to30 intervals inclusive. You may pump 10.20 liters, or 10.21, or 10.214863 liters, or any amount from zero to thirty. Using this approach, continuous data is often considered as computationally infinite. For simple record keeping, data analysts sometimes select out one number within the range to round off.

When you are collecting numerical data, you use:

- Frequencies: A frequency is a rate that one thing happens over an amount of time or at intervals in a dataset.
- Proportion: The proportion is calculated by dividing the frequency by the overall sum of events.
- Percentage.

Visualization Methods: to examine nominal information, you will be able to use a bar chart or a pie chart.

Categorical

Categorical data is often any data that is not a number, which might mean a string of text or date. These variables may be softened into nominal and ordinal values, although you would not typically see this done. Categorical data represents characteristics. So, it will represent things like a person's language, ethnicity, etc. Categorical data may also defy numerical values (Taking an example: 1 for "on" and 0 for "off"). It is important to note that these numbers do not have any meaning in the mathematical sense.

Ordinal

Ordinal values describe discretely and ordered units. Examples of ordinal values include having a priority on a virus as "Critical" or "Low" or the ranking of a restaurant as "Five Star" or "Three Star." Therefore, it is possible to encapsulate your ordinal information with frequencies, proportions, and percentages. By doing this, you will be able to visualize the data through a bar or pie chart. Besides, you will be able to use median, percentiles, mode, and also the interquartile range to compile your data.

In addition to ordinal and nominal values, there is a select sort of categorical data known as binary. Binary data types solely have two values – yes or no. They could be described in numerous ways like "True" and "False" or one and zero. Binary data is employed heavily for classification machine learning models. Samples of binary variables will include whether or not an individual has stopped their subscription service or not, or if an individual bought a car or not.

Clustering

The purpose of classification and clustering algorithms is to make sense of and extract worth from giant sets of both unstructured and structured data. When you are operating with vast volumes of disorganized data, you would be smart to partition the data into some logical groupings before trying to analyze it. A loose definition of a cluster can be the method of organizing items into groups that have members with similar characteristics.

Clustering and classification allow you to look broadly at your data en bloc, to create some logical structures supported by what you discover there before delving deeper into the practical analysis.

Clusters, in their purest form, are sets of data points that share similar characteristics, and cluster algorithms are the ways that classify these points of data into totally distinguished clusters grounded by their correlation. You will see cluster algorithms applied for disease classification in medical sciences. You will also know the algorithm applied for client classification in research and environmental engineering and health risk assessment.

There are two different clustering ways, depending on your dataset:

- **Hierarchical**: Algorithms produce individual sets of nested clusters, every in their own rank level.

- **Partitional**: Algorithms produce only one set of clusters.

Clustering may be thought of as the most significant unsupervised learning hurdle; thus, like every alternative drawback of this sort, clustering deals with finding a system in an exceeding compilation of unlabeled data.

Importance of Clustering

The goal of clustering is to work out the inherent grouping in a collection of unstructured data. However, do we decide what consists of a decent cluster? It may be shown that there is no perfect "ideal" criterion, which might be free of the ultimate goal of clustering. As a result, it is the user that should provide this criterion in a manner that the outcome of the cluster suits their desires.

Take, for instance. People tend to be interested in finding representatives for homogenized teams, in finding "natural clusters" and describe their anonymous properties ("natural" data types), to find advantageous and suitable groupings or to find uncommon data pieces.

Clustering Applications

Clustering algorithms may be applied in several industries, for instance:

- Marketing: to find groups of shoppers with similar behavior given an extensive customer information database containing their features and historic shopping records.
- City-planning: characterizing groups of homes per the type of house, price, and geographical location.
- Biology: using taxonomy, the classification of animals and plants is given their features.
- Libraries: checking books;
- Insurance: distinguishing groups of motor contract holders with a higher average claim cost; as well as identifying fraud;
- Earthquake studies: determine earthquake epicenters to spot dangerous zones;

The primary necessities that a clustering algorithmic rule ought to satisfy are:

1. Scalability
2. Ability to deal with differing kinds of attributes
3. High dimensionality
4. Discovering clusters with discretionary form;
5. Interpretability and value
6. Minimal needs for domain information to work out input criterion;
7. Capacity to handle noise and unusual data objects
8. Insensitivity to the arrangement of input records

Clustering Limitations

There are a few issues with clustering, among which include:

- Current cluster techniques do not cater to all the requirements sufficiently (and simultaneously)
- Big data and dealing with a wide range of dimensions may be problematic due to time complexity
- The results of the classification algorithm (which in several cases may be discretionary itself) may be taken in numerous ways.
- The effectiveness of the cluster strategy depends on the description of "distance" (as in distance-based clustering), but if a physical distance does not exist, we have a tendency to "define" it, which is not an easy task, particularly in multidimensional spaces

Applications of data science include:

Internet search engines use data science algorithms to deliver the matching results for search queries in a fraction of a second.

Digital Advertisements, including the complete digital marketing spectrum, uses the data science algorithms- from show banners to digital billboards. This is often the main reason for digital adverts obtaining a higher click-through rate than traditional advertisements.

Recommender systems do not merely make it simple to seek out relevant products from billions of products on the market yet also adds loads to user-experience. Many firms use this technique to market their products and suggestions per the user's demands and connection of data. The recommendations are supported by the user's previous search results.

Watson, by IBM, is an AI technology that helps physicians quickly determine critical data in a patient's medical history to gather relevant evidence and explore treatment choices. Watson collects a patient's medical records then provides its evident-based and personalized recommendation powered by data from a curated assortment of journals, textbooks, and pages of texts that offer doctors instant access to an abundance of information customized to the patient's treatment plan.

Blueberry is an example of a robot created by Kory Mathewson, that can perform improv comedy as a result of being fed subtitles from many thousands of films. He trained it to make lines of dialogue for an improv performance by rewarding it once the dialogue was sensible and punishing it once it spoke nonsense. While Blueberry will not be auditioning at a talent show in the near future, this lovely robot will often hit the proper notes with funny lines.

Chapter 8: Internet of Things

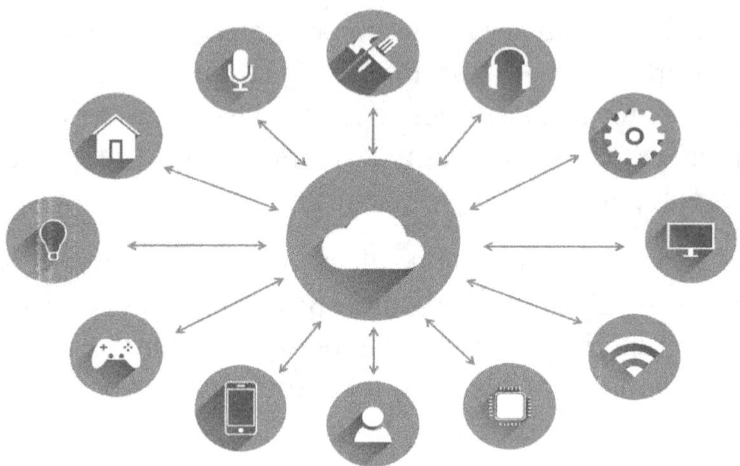

One of the most common concepts today in the mobile app development world is the Internet of Things. From the simplest consumer applications such as wearables and smart homes to the most complex industry solutions like driverless forklifts, the Internet of Things (IoT) has penetrated every aspect of technology and is gradually changing the way that we live, interact with others, and work with internet-enabled devices.

In a report recently published by Statista, more than 31 billion devices, including smartphones, cars, smartwatches, and wearable, will be connected by the end of 2020. The number is currently recorded at 23 billion, which is a demonstration of the exponential growth of IoT.

While the concept of IoT has been in the public realm for a very long time, many people are still not familiar with it or how it functions. In this chapter, we are going to understand the key concept as well as its application.

What is the Internet of Things (IoT)?

The Internet of Things (IoT) simply refers to the connected ecosystem of physical devices, appliances, vehicles, and other things that can collect data and exchange it through a wireless and wired network without any human-to-computer or human-to-human intervention. Through the enables integration and exchange of data by the physical devices, this technology focuses on enhancing people's lives through the provision of comfort and simplicity with possible efficiency.

Through the integration of complex technologies such as Artificial Intelligence (AI), machine learning, and machine-to-machine communication, IoT aims towards extending the connectivity of physical things beyond the common internet-supported devices such as tablets, PCs, and smartphones to a wider spectrum of non-internet enabled objects like coffee makers, door locks, and washing machines. This approach is important as it will enable humans to control and monitor them with the help of a simple mobile gadget.

How IoT Works

Just as other computer devices and systems have predefined components and steps, so does the Internet of Things. A complete IoT system constitutes four components that work together to ensure the system provides a desirable outcome. The key components are discussed below.

- **Sensors/Devices**

    Sensors or devices collect even the minutest data from the surrounding environment, which could include simple information as geographical location or complex data like the health history of a patient. In order to pick up this form of data, multiple sensors can be bundled together to form a device that is able to perform complex tasks than a simple sensing activity. For example, a smartphone is composed of several in-built sensors like Camera, GPS, and Accelerometer, without which could hinder the phone from sensing any data.

Thus, the first step to the functioning of IoT is the collection of minute data from the surrounding environment through the use of multiple sensors.

- **Connectivity**

After data has been collected by sensors or devices, it goes to the cloud infrastructure, which is the IoT platform through the use of a medium. Here, there are several wired and wireless networking technologies like Wi-Fi, Bluetooth, LPQAN, Cellular Networks, and Ethernet that transmit the collected data. Although the connectivity options represent the interaction of connection range, power consumption, and bandwidth, choosing the one that transmits data to the cloud significantly depends on the specific requirements and the complexity levels of the IoT application.

- **Processing of data**

After data has been transmitted to the clouds, it is stored, processed, and analyzed through the use of Big Data Analytics Engine to enhance decision making. The analysis process can be as simple as checking the temperature of an AC, or as complex as identifying burglars into a house through the use of surveillance cameras. Once the data is processed, it is used to conduct immediate action that can turn an ordinary physical device into a smart one.

- **The user interface**

  In this last step, the IoT system notifies the consumer about the action through text, email, alert, or notification sound that is triggered by the system. The user can then either leave the notified action intact, manually perform an action that affects the system, or proactively check the IoT system. For example, if a user realizes some changes in a particular room, he/she could adjust the room temperature using the IoT app that is installed on his/her device.

Benefits of IoT

Although the main goal of IoT is to automate human life and make everything efficient, there are several other benefits that the technology offer to consumers and businesses. Some of the key benefits include:

- **Easy access to data**

  Everyone, including entrepreneurs and marketers, love quality data, and with the invention of IoT, firms are able to effectively access data that relate to their consumers and products more than ever. They are able to take advantage of the real-time operation insights they obtain from IoT devices to monitor the behaviors of their customers, enhance the consumer experience, and make smarter decisions. Simply put, the more information you have, the more you are likely to make the right choices.

- **Better tracking and management**

With IoT, industries are able to track, and management has become very easy. From monitoring road traffic and weather changes and tracking inventories to notifying authorities of any illegal or worrying activities, IoT has revolutionized the way that people manage and track their business assets. IoT system is not just about smart devices or smart homes; it is now a smart office, a smart service, and smart everything.

- **Efficient utilization of resources**

Whether it is home, hotel, office, or car, the Internet of Things facilitates the efficient utilization of assets with the aim of improving productivity. Through the advantage of interactions of machines, an IoT system is able to gather real-time data through the use of actuators and sensors in order to further use the big data to improve efficiency while also minimizing human intervention. An example is when your home appliance alerts you of a completed task, you will never worry about the efficient use of power at your home.

- **Automation and control**

  IoT promotes automation since the devices are always connected to each other via a wireless network, which makes them operate on their own without excess intervention. For example, your home appliances like washing machines, air conditioners, and an oven can operate automatically even without your monitoring or controlling them remotely.

- **Enhances comfort and convenience**

  We are living in a society that is fast-paced, and people are very busy that they don't care about little things such as reading power meters or switching off lights. However, with IoT, such problems can be addressed since the interconnectivity of data and devices offer full control over devices that are connected to the IoT system. With your ability to control your home devices through a centralized device such as a smartphone, you can enhance convenience and comfort.

- **Saves time and money**

  The use of IoT involves getting more done with as little effort as possible through automated tasks and little human intervention. Being able to accomplish your cumbersome tasks faster without the use of energy, IoT enables you to save your hard-earned money and save your quality time. For instance, if one of the electronic appliances is able to turn itself off after a task, you will save the effort and time you may have required to witch the appliance of manually.

Use Cases of Internet of Things

Some of the real-world applications of IoT are:

- **Smart Home**

  A smart home remains the best example of IoT technology. Smart home devices and systems offer optimum security and convenience to users and are designed to save energy and time. With a smart home, you can control everything in your home, from temperature to lighting with just a simple smartphone control.

- **Wearables**

  Although there are wide varieties of smartwatches and fitness tracking devices in the market, many superpowers like Intel and Samsung have begun investing in IoT-powered wearables. Through the use of installed software and sensors, these devices are able to track and monitor significant health metrics like eating habits, blood pressure, sleeping habits, heart rates, and caloric intake.

- **Smart cities**

  Different countries, including the UK, Spain, Japan, and South Korea, are making efforts to colonize smart cities with the aim of providing their citizens with the best and healthier environment to live in. The countries collect data from assets, citizens, and devices with the aim of solving major challenges that people in those cities face like crime, congestion, water distribution, and waste management.

- **Automotive and transportation**

With IoT, companies in the automotive industry like Tesla, Ford, Volvo, and BMW are already planning to enhance their in-car experience. With advanced technologies such as computer vision, sensors, sonar, internet, and maps, these vehicles operate without drivers and can run without any human assistance. If combined with machine learning, IoT in the transport sector can help to perform roles like smart traffic control, smart parking, fleet management, and logistics.

- **Medical and Healthcare**

The field of medicine is currently utilizing IoT powered devices to offer remote emergency alerts and health monitoring services. Through the use of smart healthcare devices, doctors are able to monitor their patients' health away from the clinical environment and provide medicine on the basis of relevant data. Also, doctors in the emergency department keep themselves ready for any emergencies because IoT enables them to be aware of their patients' medical conditions.

- **Industrial IoT**

  Industrial IoT is the subfield of IoT that leverages the technology used to solve industrial problems, eliminate efficiencies, and automate industrial processes. Apart from the manufacturing industry, other applications of industrial IoT include aerospace, energy management, defense, and futuristic farming.

Internet of Everything and its Relation to IoT

The concept of the Internet of Everything (IoE) is based on the ideology of all-rounded intelligence, connectivity, and cognition. The concept refers to the intelligent internet connections that are not restricted by any devices such as smartphones, computers, and tablets. IoE encompasses any object with digital features that is able to connect to the network of other objects, processes, or people to generate and exchange information as well as facilitate decision-making.

The philosophy of IoE depicts a society where billions of sensors are implanted in different machines, devices, and ordinary objects in order to expand their networking opportunities, thus enhancing their ability to be smart. The key features of IoE include:

- **Decentralization**—With IoE, data is processed in several distributed nodes rather than in a single center.

- **Data input and output**—IoE enables external data to be put into devices and shared with other network components

- **Relates to all digital transformation processes**—IoE technology is interconnected with several digital processes, including AI, Big Data, machine learning, fog computing, and cloud computing.

Elements of IoE

There are four constituent elements of IoE, which include:

- **People**— People offer personal insights through applications, websites, and connected devices that they use. This data is usually analyzed by smart technologies and AI algorithms in order to understand the issues affecting humans and provide relevant content in line with personal needs. This helps to provide quicker solutions necessary for making decisions.

- **Things**— These are the pure concepts of IoE. Several physical items that are embedded with actuators and sensors help generate data about their status and send them to specified destinations across the IoE network.

- **Data**— Raw data that is generated by devices does not have any value. However, once it is summarized and analyzed, it becomes so priceless that it can empower several systems and intelligent solutions.

- **Processes**—Different processes involved in IoE ensure that the correct information is received at the right time and sent to the right person. The goal of IoE processes is to ensure the best usage of Big Data.

The Relationship Between IoT and IoE

The Internet of Things is not the same as IoE. The core difference between the two lies in the number of pillars that each concept has:

- IoE relies on four components, which include things, people, data, and processes

- IoT only focuses on physical objects

In essence, IoT involves the interconnectivity of objects that send and receive information while IoE is a term that widely includes several technologies and people as the receiving end.

Although the two concepts differ, they have common similarities:

- **Decentralization**— Both IoE and IoT are distributed and lack central points; each of them work as a small management center focused on performing certain tasks independently from human intervention.

- **Security issues**— The systems are still very vulnerable to cyberattacks and penetration; with more connection of devices, they become more susceptible to breaches.

How AI and IoT Can Be Combined for More Success and Reliability

While IoT collects data from the surrounding environment, the AI acts as the master "brain" that analyzes the data to aid in decision making. This means that IoT collects data while AI activities process this data and provide its meaning. The useful integration of the two technologies is more apparent in the current systems used by sport tracking devices such as Siri, Alexa, and Google Home.

With connected devices, thanks to IoT, more data can be collected to offer incredible knowledge for companies and the general public. However, this data cannot be helpful to anyone if it is not analyzed. And this is where AI comes into play, utilizing the huge amounts of data to predict trends and offer solutions. Thus, technology developers must improve the integration of AI with IoT data through the following:

IoT and AI Data Analysis

AI relies on four different IoT data analyses, which include:

- Streaming data visualization— In order to integrate IoT with AI, developers must treat streaming data by displaying, defining, and discovering the collected data using intelligent ways to enhance decision-making.

- Accuracy of data time series—It is necessary to maintain a high confidence level of data that has been collected to enhance the integrity

- Advanced and predictive analysis—It is very important to analyze data collected using advanced and predictive techniques.

- Maintain the flow of geospatial and real-time data.

AI in IoT Applications

Visual microdata gathered by IoT can only be understood with AI applications, which would be able to interpret the context of the images. New sensors will also enable computers to collect data through audio formats within the user's environment.

Chapter 9: Artificial Intelligence Superpowers

The United States has protracted its stay as a frontrunner in synthetic intelligence. But conferring to Dr. Kai Fu Lee-one of the sphere's utmost cherished whizzes on AI, China has at present trapped up with the United States at an astonishingly fast stride. As the rivalry between the two countries rises, Lee is foreseeing the two countries, China and the United States, bring into being a foremost duopoly in AI. The Chinese and American administrations will have to familiarize themselves with the fluctuating commercial scenery as a result of AI disruptions.

China and the United States, by this time, have a prime above other nations in artificial intelligence engineering. Even though you will realize improvements and innovations from other nations such as the United Kingdom, France, Singapore, Canada, these countries are dedicating a lot of resources in the event of AI. They have already established active AI inquiry labs manned with abundant ability, but they a shortage in the venture -investment ecological unit and enormous employer stations to produce the Statistics that will be key to the era of putting into practice that you find in the United States and China.

As AI establishments in the United States and China gather supplementary statistics and endowment, the honorable sphere of data-driven progress in broadening their tip to a point somewhere its backbone convert firm for the other countries to launch any meaningful challenge. China and the United States are at this time nurturing the AI monsters that will take over worldwide markets and excerpt treasure from clients everywhere in the earth.

Lee explains that China has precipitously gathered up to the United States at a surprisingly fast and unpredicted, and as a result of this competition, he argues, dramatic changes may happen sooner than you expected.

Supreme whizzes have, by this time, foreseen that AI will have an upsetting influence on blue color occupations. The experts have urged the two nations to agree to take and encircle the abundant accountabilities that emanate with substantial technical power. Lee argues that it is not only the blue-collar jobs that will be affected. He believes the technological advancement will have an impact on white-collar jobs as well. Lee goes ahead to provide a perfect depiction of the tasks which will be exaggerated and how rapid. He gives his suggestion on the resolutions to some of the philosophical deviations in times gone by that are upcoming quickly.

Who Will Triumph the AI Competition Between the US and China?

Currently, in internet AI, Chinese and American corporations are on equivalent footing. However, it is, however, foreseen that Chinese machinery syndicates will have a minor improvement over their American counterparts in the next five years to come.

In business AI, the states are leading the pack. However, China is expected to close in fast in the next five years. It is likely that China will hold up behind in the commercial world, but it is likely to tip in unrestricted facilities and businesses with the prospective to swell in advance on old-fashioned schemes.

Awareness AI is also outcoming its manner into our day-to-day subsists. It is digitizing your bodily biosphere, erudition to be familiar with appearances, appreciates applications, and has graphics of the biosphere around us. The Chinese culture has little concern about statistics disclosure will provide a positive edge in the carrying out. So, consequently, at this time, China's is slightly ahead of the states in perception AI but is expected to open a wide range and lead the rules and the rest of the biosphere in discernment AI in the next five years.

In Autonomous AI currently, the United States is leading the rest of the world in self-driving cars, but as advancement in autonomous technology continues to be realized, China is expected to level the completion with the United States in the next five years. China already has the authority in hardware concentrated solicitations such as self-governing drones.

Is China the Succeeding Global force?

China's general attention on AI and the possessions it is conveying to the struggle could let it take the lead in the field of AI, according to AI experts. The United States has long been perceived as the universal spearhead in modernization, which comprises the field of artificial brainpower. China has extensively been regarded as a technical impersonator. This, nevertheless, has changed drastically and may not be the same anymore. According to the experts, China is poised to take the lead in the coming few years. China's favorable policies on AI, its enormous group of statistics, and the gigantic marketplace, as well as the existence of painstaking and go-getting industrialists, could support the nation overhaul the United States in Artificial intelligence.

However, it is essential to note that the two countries are parallel universes, and each is making its own progress in the world of AI. The United States is a quiet way forward in the primary expertise from the inquiry test center and campuses. But China is nowadays enchanting the tip in putting into practice and constructing worth by means of artificial brainpower across all solicitations and diligences.

The skillful application of AI may help China to catch up and surpass the United States. According to Lee, this will also support folk's re-experience what it means to be humanoid. He believes that the whole job market will drastically change. You will be using more of AI human brainpower. These are precise AI engines that explain communal hitches at a stretch. For example, drivers that can make loan decisions foe banks, or robotics which can perform chores like washing dishes or fruit picking. Although this could mean some human jobs will be replaced, AI will be good at producing tools for creative and professionals.

Factors Helping China to Become an AI Superpower Quickly

The Chinese government is determined to lead the world in artificial intelligence by the year 2030. Experts tend to agree with China's ambitious goals. It is expected that by 2020, China would have caught up with the united states in AI. Experts also believe that if they maintain their trajectory, China would be far ahead of the United States by the year 2025, and by the year 2030, China will be dominating the industries of AI.

With a GDP of $14 trillion, China is projected to a justification for above 35% of the ecosphere's commercial progress from 2017-2019. This is almost double the United States' GDP prediction of 18%. China's impressive growth is credited to its use of AI across the major industries.

PricewaterhouseCoopers has projected that AI's deployment in strategic industries swill help adds an additional $15 trillion to the international GDP using China captivating a big chunk of that at $7 trillion. North America is expected to take home a paltry $3.7 trillion in expansions. In 2017, China accounted for 48% of the biosphere's overall start-up backing as equated to the united states' 38%.

By this time, the Chinese money in AI imperfections and self-directed automobiles have extended $300 billion, with Chinese companies like Alibaba pledging to invest a lot of resources in international research labs in the United States and Israel.

According to Kai-Fu, there are four main factors which have helped China to move very fast to the top list of the world AI superpowers:

Availability of Data

China is taking advantage of its significant quantity of data to make great leaps in Artificial intelligence. China's Tencent

WeChat stage without help has above one billion on the go recurrent consumers. This surpasses the combined population of Europe. When it comes to mobile spending, China beats the United States by a proportion of 50:1. Furthermore, Chinese disbursements on e-commerce are just about double that of the united states.

China is also witnessing an explosion of online to offline startup us, which are helping in making more data available. Another key advantage of Chinese Data is that they concentrate all their data in one place. Chinese AI businesses like Tencent have generated a combined operational network, while Americas' data tends to be spread across various platforms. A key example is American overpayment and transportation data, which are fragmented across multiple platforms.

If you compare mobile payment data between the two countries, you get exciting facts. While the United States saw $112 billion substance of portable disbursements in 2016, Chinese portable disbursements situated over 49 trillion in the similar time. This means AI developers can mine crucial data from Chinese mobile payments like WeChat Wallet and Alipay. This permits them to produce maps monitoring hundreds of millions of operators every single interchange.

With the upswing of bike-sharing startups like Chinas' ofo and Mobike, Chinese syndicates can construct the use of profoundly surfaced maps of inhabitants association. This permits them to understand the whole thing from your operational practices to your spending mundane.

And as the Chinese's AI facemask acknowledgment aptitudes progress, the maps are gradually being uninhabited with appearances whether you are available or disconnected.

Chinese AI syndicates are assimilated operators' online performances with their bodily biosphere. And as they do so, the Data they collect gives them a significant advantage over their competitors in the United States.

AI Expertise

Although China is still new to the United States in artificial intelligence, china's AI researcher has caught up with the states. When deep learning in the United States made a successful breakthrough in 2012, China was in its initial stages in the AI revolution.

But while AI researchers are still located mainly in the United States, be in support of something businesses like Google, Chinese tech enterprises are rapidly concluding in. At this time, in Academia, Chinese Artificial brainpower scholars are at the same level as those of the united states. For example, an equivalent sum of conventional papers originated from the nations and China in the recent 2017 AAAI conference.

There also has been a noted increase in the partnership amongst china's top tech corporations and developing undergraduate flair. For example, tech firm Tencent has been sponsoring students at an AI lab in Hong Kong's school of science and expertise. The students are also granted access to billions of WeChat data. Moreover, leading Chinese tech firms like Baidu, Didi, and Tencent have all established their specific investigation firms.

As a result of these investments, AI Chinese companies are now leading the world in some technologies. For example, china's appearance++ these days tops the biosphere in appearance and duplicate acknowledgment AI. They beat

American companies like Google and Microsoft and Facebook in the 2017 COCO appearance acknowledgment antagonism.

Chinese's speech acknowledgment corporation iFlyTek has outcompeted America voice recognition companies such as Alphabet's DeepMind, Facebook, and IBM Watson in likely linguistic dispensation.

China's Aggressive Entrepreneurs

China's impersonator age saw a big movement of shoddy goods and fake impression. However, this ear helped to inaugurate approximately the utmost inexpensive and hostile industrialists in the biosphere. In fact, China is into inconceivable tough work. Corporations work 9 am to 9 pm. six to seven spells a week shorn of exclusion. Businesspersons are also hard top-down with a solo individual creating all pronouncements. So, the choices are swift. It is all about moving on and executing.

These Chinese tech entrepreneurs have been aggressive in their pursuit to beat the competition. They have devoted a lot of their time and resources to outpace and outsmart parallel startups from elsewhere in the world. It has been noted that the swiftness of the work is abundant quicker in China than in any additional portion of the tech world, like in Silicon Valley. Business opportunities in China are snapped at very fast as compared to the time it takes for tech opportunities to be identified and considered in the United States.

Today, china's AI proficiency has derived from time and startups partake educated to tailor-make American impersonator merchandise to garb the Chinese consumer necessities. These businesspersons are succeeding in getting rid of the copycat tags and are building businesses with no analogs in the United States and Europe.

China has also produced the world's leading AI companies like Baidu, Alibaba, and Tencent. It is also making a significant contribution to the establishment of the world's most valuable AI start upon, such as the Chinese computer startup SenseTime. The startup is currently the most valuable AI start-up in the world. The AI startup product has features that make it possible for your face to be identified as well as your age. It can also determine your impending procuring routines. SenseTime is, nowadays, foremost the biosphere in face pack appreciation machinery smearing their AI to the whole thing from transportation investigation to worker endorsement.

China is also home to over 160 unicorns valued at over $600 billion. China is using its growing expertise to advance further their AI startup agenda.

China's Favorable Government Policies

The Chinese government has issued plans to mark China as the worldwide epicenter of AI modernization with the aim of 1 trillion RMB AI industry by the year 2020. This is an equivalent of $ 150 billion USD.

After the announcement of these grand plans, the Chinese VC investors have invested huge sums into AI startups. Besides, the chines government has been spending a lot of money on their STEM research, which has seen it grow by double digits in recent years.

Besides, China's political system has engaged in recreation, a vital character in the progression of AI research. Native bureaucrats are incentivized to outcompete others in CPP advantages. Because of this arrangement, each leader is striving to win over AI businesses and businesspersons with substantial subventions and satisfactory guidelines.

Mayors transversely the nation have assembled out modernism zone, incubators, and government-backed VC monies. All are unfluctuating paying for expenses of researchers such as rent. This has played a favorable environment for AI startups

Moreover, China is set to capitalize $2 billion in AI expansion park. The park will stock up to 400 AI initiatives and a nationwide AI lab, copyrights, communal invention, and a driving R&D. Besides, the province of Hangzhou has also propelled its personal AI park with a deposit of $1.6 billion USD. There are other cities and regions totaling 19, which are investing a lot of resources on AI is driven city infrastructure as well as on policy development.

Cities like Xiongan Innovative Zone are constructing all-inclusive AI cities in the succeeding two periods. The cities majorly focus on the development of self-directed automobiles, solar panel entrenched infrastructures, and CPU vision geared set-up.

In addition, China's local governments are collaborating with the country's prominent AI tech corporations to derive up with business facilities. As an outcome, corporations like Baidu, Alibaba, Tencent, and iFlyTek are accompanying with countrywide administrations like China's National Engineering Lab for Subterranean Scholarship Skills to invent AI research.

The government has also made available funds for ai innovation and development. The Chinese startups and established tech firms are making use of these funds to grow further and develop new AI products. China's lead in AI appears well established with the help of lavishness of straightforwardly reachable administration funds, smooth infrastructural refurbishments, prominent AI inquiries, and aggressive entrepreneurs base.

As a result of all these factors, the usage of AI in China is no slower enigmatic. AI is no further extensive views as very advanced technology that few enjoy. AI is a nowadays vulnerable cradle, and new alumni from the institution are expected to start using AI engineering to build AI products in a few years to come.

China's New AI Superpower Status, Its Implications, And the New Order

Subsequently closely four eras as the industrial unit of the biosphere, China nowadays is treading into the unique characters in the worldwide budget. The country is considered the leading pivot for pioneering submissions of artificial brainpower. Conferring to one current tale, by PricewaterhouseCoopers, of the $15.7 trillion in international treasure AI is anticipated to breed by the year 2030, China is expected to contribute $7 trillion alone.

It is expected that in the coming days, China's world-class entrepreneurs will apply deep learning to any problem that promises huge profits across all the industries. They are being aided by Silicon Valley's weakness at unwillingness and resistance to localization. As compared to the earlier internet services, AI has a much higher localization quotient. As a result, every divergence between Chinese preference and the standard global product will become an opening which local competitors will attack.

Social Implications

The rise of China as the world's AI superpower could come with several implications. There can occur a large-scale communal complaint and politically aware downfall initiated by pervasive joblessness and wide-open difference. In the coming days, AI will have great potential to disrupt and destroy lives with the expected impact on labor markets and social systems.

The emerging platforms leveraging an AI foundation have a natural affinity for a winner take all economics eroding the competitive mechanisms of markets in the process. This is likely to occur across all industries with a skill bias that divides the job market and squeezes out the middle class.

Economic Implications

Moreover, inequality is expected to rise as a higher meditation of capital gets in the influences of a few. The gap amongst the rich and the unfortunate will remain to grow at an individual level as well as at country levels. As the robot operated factories continue to move closer to the markets, the ladder which the developing countries use to climb out of Poverty will be cut off.

There will be a widespread struggle because people take part been trained to develop a sagacity of value from functioning. The intensification of AI will encounter these standards and intimidate them into weakening this logic of lifetime tenacity.

Implications of Jobs

The rise of AI will definitely have a negative impact on jobs. Within 15 years, AI is estimated to be capable of eliminating over 50% of US jobs as a result of one to one replacement and ground-up disruptions.

In addition, existing high paying professions like medicine will take a different path to the same end. As intelligent machines surpass human's ability to diagnose diseases and recommend treatments, the doctor's role will be significantly limited.

Implications of AI on Global Peace

Military planners both in America and China conquer that AI makes war more likely. Autonomous systems may create a conflict waged by robots less costly in terms of human casualties, but the systems also make conflict more likely because AI will be deciding the speed and the course of the war. China's military and political leaders believe that by 2025, autonomous weapons will run the battlefield with minimal human engagement. It is very easy to see why such prospects could easily destabilize the fragile world peace.

Conclusion

Thank you for making it through to the end of *Artificial Intelligence for Beginners: Easy to Understand Guide of AI, Data Science, and Internet of Things. How to Use AI in Practice? Revelations of AI Superpowers Explained for the Real World.* Let's hope that it was informational and was able to provide you with the basic knowledge you need to understand the concepts of Artificial Intelligence and related technology. By finishing this book, you will be able to possess the mastery that you seek in understanding the role of artificial intelligence in influencing human behaviors, and its impact on the career market.

We have gone through the definition, goals, and types of artificial intelligence and its relationship with emerging technologies, including robotics and the Internet of Things (IoT). This book has offered easy-to-use but very powerful and effective definition of concepts that are crucial in understanding artificial intelligence. It provides a great overview of how the world is gradually adapting to technological changes with the aid of artificial intelligence. You are now familiar with the relationship between AI and robotics and IoT, as well as the possible differences. You have also learned that almost every aspect of our careers is gradually embracing AI for efficiency.

Having understood the key concepts of AI, its goals, and how it works, the next thing you would want to do is to decide to invest in the field due to the vast opportunities available. Currently, China is at the top in the investment on AI-related technology with mega giants such as Google and Amazon following suit. With the knowledge provided in the book, you can learn the more opportunities that exist in such technology.

Finally, if you found this book useful in any way, a review on Amazon is always appreciated!

Machine Learning for Beginners

Easy Guide of ML, Deep Learning, data analytics and Cyber Security in practice. Modern approach of Neural Networks, Predictive Modeling and Data Mining with 50 Key Terms

INTRODUCTION

Congratulations on downloading *Machine Learning for Beginners,* and thank you for doing so. The future is predicted to utilize machine learning techniques to solve everyday situations, and downloading this book is the firsts step to act before it caught up with you. Since its incorporation in the 1950s, machine learning has a broad and complex concept which may become more challenging for beginners to venture. However, the initial step in understanding machine learning is usually the easiest and among those found in the following chapters.

The following chapters will, therefore, highlight the primary aspects of machine learning, which primarily the building blocks are. Beginning with the basics provides you with a clear idea of what machine learning is all about. Consequently, machine learning uses multiple terms which may become challenging or sometimes misunderstood, especially for beginners. In this case, the chapters will isolate these vocabularies while providing detailed information about what they stand for when it comes to machine learning.

When you are quite familiar with what machine learning is all about, including basics and terms used, the book will highlight the introduction to different concepts. As the book is primarily for beginners, understanding basics, as well as a brief and detailed introduction to different areas of machine learning, is essential. Besides, you will learn about the dos and don'ts of machine learning, algorithms of machine learning, and fundamental theories in this topic.

There are plenty of books on this subject on the market, thanks again for choosing this one! Every effort was made to ensure it is full of as much useful information as possible, please enjoy!

CHAPTER 1: FUNDAMENTAL CONCEPTS OF MACHINE LEARNING

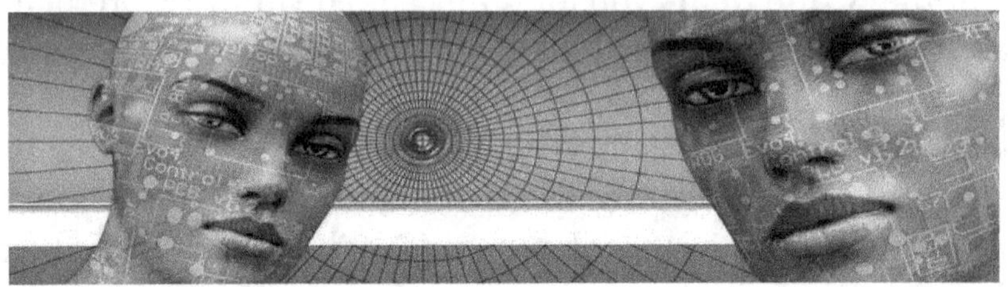

"Machine learning is the science of getting computers to learn without being explicitly programmed"
Sebastian Thrun

Basics of Machine Learning

Machine learning is described as the ability for machines to understand human behavior and speech autonomously from experience, analysis, and observation tessellation based on specific data without programming. When compared to machine learning, programming entails writing a set of given codes or programs and inputting explicit instruction where the computer or machine follows to perform a given task. Whereas, machine learning depends on a data set where the device readily identifies and analyzes the data set and make decisions without any external assistance.

These machines would, therefore, perform a given task automatically when they observe and learn from the dataset without being fed more similar instructions. Machine learning covers a broader field of computing. As such, beginners should expect several courses or articles to cover these topics fully. For example, Facebook uses machine learning to such as facial recognition algorithm to quickly identify an individual; therefore, users can tag photos when uploaded. Another example is Google Voice Service used in multiple smartphones where people can readily use voice search assistance.

How Machines Learn

Machines have been made to observe, learn, and analyze different functions and events like how humans do. The first concept of machine learning began in 1943, where neural networks of the human brain were introduced to computers. However, the development of computing devices, both hardware, and software, at the time was slow and solely used in large organizations and academic institutions. More so, acquiring the necessary data was difficult, unlike the current days where we quickly access the internet and get what we need.

Humans acquire the knowledge about a given thing, memorize it and retrieve it in the time of need when a similar occurrence occurs in the future. This behavior is conducted by a network of neurons in the brain. The same pattern was introduced to machines making them work like how brains recognize different features and distinguish them at different times in the future when they arise. Machine learning can also take various forms when computers learn. That is, some may become specific and perform a given task while others are general learning multiple topics and implementing them like how our brains work.

Basic Terms Commonly Used in Machine Learning

Bias

Bias is a term used in machine learning to describe the model developed either if the predictability level is low or high. That is, if the bias is low, then the machine has a much higher predictability level with an increased biasness suggesting a lower predictability level. Therefore, biasness in machine learning tends to determine the level of mistakes made when executing datasets. The primary benefit of bias in machine learning is that it readily provides a comparison of algorithms between two machines if the problems are alike.

Cross-Validation Bias

This is a term used to suggest a technique used in machine learning to measure the accuracy of performance of the model. Cross-validation bias usually provides the performance relatively closer to the intended outcome, primarily when the model is used in the future without the primary dataset. This technique utilizes the present knowledge fed to predict future unseen situations. When provided with the correct model, cross-validation bias can result in much higher performance.

Underfitting and Overfitting

Underfitting is a term typically used to describe events where the machine learning models are unable to predict the intended dataset with the required level of accuracy. There are several reasons which may result in these situations, for instance, when you fail to collect the right features or on complex problem statements. On the other hand, overfitting is when the model fits excessively or becomes too complicated also used in statistics. Here, the machine learns in detail and much more profound than expected; therefore, the outcome, though positive, may impact the performance negatively.

Deep Learning, Machine Learning, and Artificial Learning

Beginners may face a challenge when trying to differentiate between machines, artificial and deep learning when it comes to machine learning. Artificial learning comprises of tools which simulate how we behave and make decisions intelligently like us. Machine learning, as mentioned, is the ability of machines to learn automatically without programming using codes or programs. Deep learning combines both artificial and machine learning but with the assistance of algorithms and artificial neural networks.

Components of Machine Learning

Data Collection and Preparation

Machines can never work unless fed with the relevant data which triggers the application of the intended task. The first step in machine learning is collecting and preparing data which is divided into training and testing data. An example is when you want to create software which quickly identifies photos of specific persons. First, collect the data of the people, intended and set your parameters based on age, sex, or other relevant information and input your data on a model which can be tested.

Selection and Training of Models

Choosing your model and training it depending on your requirements as another essential component which determines the outcome of your machine learning strategy. However, several algorithms and models of machine learning incorporated have undergone modification to provide solutions to problems at hand. Choose a model which can be readily trained to fit your criterion for your machine to understand and provide solutions immediately.

Model Evaluation

When a given data is fed into a machine, it readily learns features and patterns of the data train itself to work like how to learn new things. More so, when fed a well-tested data, machines can produce excellent results when deciding on their own. As to achieve this, work on the training data which will eventually create a model for your computer. The model would provide an algorithm where the device will follow, therefore providing higher chances of success. Then test the data to ascertain its probability of delivering excellent results at the end.

Types of Machine Learning Algorithms

Supervised Algorithms

Supervised machine learning algorithms are data sets which consist of input parameters and intended outcome labeled within the machine. Here, the machine classifies data sets into labels, and the training data included according to the given parameters. The features and patterns will be readily read and classified as intended depending on the learning capabilities of the machine. This type of machine learning algorithm can further be classified into two forms, classification and regression algorithms.

Classification algorithms are primarily responsible for classifying data into a specific label or categories with one of the most commonly used being the K-Nearest Neighbor classification algorithm. The K-Nearest classification algorithm is primarily used for classifying data depending on the similarities between different data sets. Regression algorithms, on the other hand, are specifically essential in the determination of mathematical relationship and dependency between variables. Regression algorithms can, therefore, be used to predict the outcome and also include two forms, linear and logistic recession, which determine the equations used.

Unsupervised Algorithms

Unlike supervised machine learning algorithms, unsupervised algorithms entail unlabeled data sets classified in the form of similarities of variables given. Some of the most used unsupervised algorithms include K-means clustering, recurrent, and artificial neural network. The K-means clustering machine learning algorithm entails formations of clusters or groups of similar and related data sets. Artificial neural networks are successive artificial neurons identical to that of the brain connected with nodes for the benefit of making machines think and operate as humans. While recurrent neural networks are a type of artificial neural network which uses a memory connected to its nodes for analyzing sequential data.

Reinforcement Algorithms

Among the types of machine learning is the reinforced machine learning algorithm, which makes a machine to determine ideas specifically within a given context. This algorithm is quite beneficial, especially when it comes to maximizing the outcome though it also takes risks as chances of punishments are extensive. In this case, the machine becomes aware of the mistake and make corrections; hence, most of the time, the results will become positive. Besides, you can readily program or modify the outcome to either be long-term or short-term under the Markov Decision Process.

CHAPTER 2: APPLICATION OF MACHINE LEARNING

"We need to be vigilant about how we design and train these machine-learning systems, or we will see ingrained forms of bias built into the artificial intelligence of the future"
Kate Crawford

Understanding about machine learning has crucial benefits that sitting on your computer and creating dataset models. Since its introduction into the computing world, machine learning has gained much popularity among developers and users, therefore, applied in different areas. Machine learning application range from small technological devices to more large machinery used commercially. Some of the areas used include social media such as Facebook, Twitter, and Instagram where it used for the analysis of sentiments and filtering spam, as well as in the transport sector especially in safety monitoring and traffic control. Others areas include e-commerce, health, financial services, visual assistant, and trading mainly on Algorithm Trading.

How Machine Learning Can Improve Our Society

Machine learning has a lot of potentials to impact the everyday life of humans, therefore essential in building our societies with the provision of solutions to recurring challenges. Despite being seen as a substitute to most of the human doings, machine learning accompanies different ways of how society is predicted to have developed by use of these computer learning abilities.

Healthcare

As already seen in most developed and some developing countries, the use of modem techniques of providing treatment has grown significantly. For instance, microscopic and challenging diagnosis of certain conditions may become a challenge for most doctors. In this case, the use of machine learning algorithms has the potential to learn how spotting abnormalities during analysis is done and be able to utilize the same knowledge to simplify this problem. As such, machines can be fed with models which relate to health specialists techniques and use the experience to readily impact the society positively and provide a solution to small and complicated medical conditions.

Transport

The world today is full of vehicles which are filling our highways and demand for the control for the benefit of allowing for safety and prevention of traffics. Therefore, the transport sector has, to some extent, already utilizing machine learning on the roads to guarantee safety and control of traffic. For instance, we have seen sensors on roads and rails to allow the smooth running of the sector. Besides, today, most cars, buses, and trucks use computers for different purposes, including navigation. The same has been adopted in ships, airplanes, and trains used around the world. Self-driving is also on the rise with vehicles being able to maneuver without the need of a steering wheel or someone to manage vehicle, train, ship or plane controls.

Education

VR and AR have been adopted in teaching learners about different aspects without necessarily leaving the classroom. For instance, children can easily explore the world using digital maps while in one room, therefore, making them understand different countries without actually visiting them. In the near future, scientists predict that machine learning will play a significant role in providing education to children, especially when using machines controlled by a teacher to individual students. As such, machine learning will ensure individual students are curated with courses where they are struggling henceforth promoting their academic status.

Retail

Machine learning has already played a role in the e-commerce sector, and that is expected to grow and promote the retail industry. Provision of meaningful data and insights of a given business, especially for customers, will readily encourage the growth of retail stores. Despite some suggesting that the retail sector will die due to the rise of machine learning, others oppose and indicate that will flourish with technologies like facial recognition as well as personalized customer experiences in stores.

Entertainment

With no doubt, the entertainment sector has already witnessed the benefits of machine learning despite not in the intended levels. For example, new technologies such as lights, sounds, and motions created by CGI advancements have significantly impacted the sector. Movies have, therefore showed remarkable benefit from machine learning. The same has been seen in sports, music and gaming industries with the benefit of making these devices and enjoyable moments more reliable, ease of access and effective. Machines learning are, therefore expected to revolutionize the entertainment sector and benefit the general community.

When to Avoid the Use of Machine Learning

Machine learning is a broad technological aspect which may also become very dangerous when misused. That is, when you feed your machine with irrelevant data, the chances are that the device would both learn profoundly and explore the unwanted facts or rather fail to perform. For example, a machine may opt to learn about war and eventually become too bias and learn more about killings and destruction. When such instances happen, it may simulate the same and engage in sabotage, therefore, leading to disaster.

Some of the instances to avoid the introduction of machine learning include planning and creation of datasets for a machine without data scientists, beginning the process with irrelevant data, poor infrastructure of machine learning and implementation of real datasets without proper strategy. Other cases may include mistakes on your objectives, splitting data incorrectly, and lack of modification on hidden variables. There are other significant mistakes which can be found before or during the development of machine learning procedures. In this case, you should avoid the use of machine learning, especially when implementing it to large populations.

Paradigms of Machine Learning

Supervised Machine Learning

Like mentioned above, supervised machine learning entails learning techniques of machines but within given parameters, more so, supervised learning. The input and output variables are provided a mapping to show how they relate, therefore having a managed outcome. Supervised machine learning offers the ability to readily approximate mapping functions too quickly so that new data can be inserted without altering but predicting the output variable. Supervised learning is quite like a teacher who supervises the learning of students, and any answers provided can be moderated or iterated by the teacher.

Supervised learning usually remains active until when the whole process stops after achieving the intended level of performance. Supervised learning is classified into classification learning where the problems are categorized based on similarities, and regression learning is when the output is a real value. Common examples of machine learning include linear regression, random forest, and support vector machines.

Unsupervised Machine Learning

The same has also been mentioned where unsupervised learning comprises of input data only without the probability of predicting output variables. The primary goal of unsupervised learning is to enable modeling of the structure of the dataset and learn more about the outcome. Here, the machine has no correct answer or a teacher like in supervised learning. Algorithms, therefore, work on their own to evaluate and provide rewards for their analysis. Unsupervised can be classified into clustering learning where you determine a natural grouping of data and association learning to discover relations between big data.

Reinforcement Learning

Reinforcement learning is used in several disciplines apart from machine learning more so in operation research, where it is referred to as approximate dynamic programming. In machine learning, reinforcement learning is mainly concerned with software agents and how their impacts affect the outcome of cumulative reward. It is among the three paradigms of machine learning bit differs from supervised and unsupervised machine learning. Reinforcement is usually used in machine learning to maximize the outcome of a given dataset.

An example of reinforcement learning is a puzzle where an agent has to go through different paths with an objective of finding a free means of reaching the reward without meeting obstacles. The machines, in this case as the agent, learn how to maneuver while avoiding obstacles, therefore, reaching the reward. That is, the machine will try different means with the right moves giving it a positive reward and a wrong or meeting an obstacle offering a negative reward. Therefore, reinforcement learning entails input, output, training, and continuous learning of the machine.

Data Science and Machine Learning

Data science is a multidisciplinary field of study which encompasses data inference, technology, algorithm development as well as scientific methods and systems to generate knowledge and insights useful in the solving analytical sophisticated structural and unstructured data problems. The primary goal of data science is determining findings from the general data with an effort of providing solutions to daily occurrences in different fields of science. The discovered data insights are usually quantitative essential for decision making businesses as well as data products which include algorithm solutions and operational scale.

Data science and machine learning have significant interaction, especially machine learning, which provides vital methodologies to data science. In this case, machine learning develops algorithms used by data scientists to create more for the creation of knowledge and insights for solving problems. On the other hand, machine learning significantly depends on data science which has the necessary data used to create algorithms essential for the machine to learn and work separately. Therefore, both data science and machine learning depend on each other, more so in regression, supervised clustering, and Naïve Bayes.

Machine Learning Algorithms

Linear Regression

This is a type of supervised learning which involves both input and output variables being determined. In a linear regression algorithm, the relationship between variables is expressed in the form of an equation, that is, $y = a + bx$. The primary function of linear regression is, therefore, determining the value of a and b provided that y and x are known. Linear regression algorithms are used in statistics, especially in population census and measuring of rainfall in cm.

Logistics Regression Algorithm

This is another type of supervised machine learning where the input variable is known, but the output remains unknown. The prediction is usually discrete, unlike a linear regression algorithm, more so when introduced with a transformational function. Logistics regression is quite suited in a binary classification where datasets take the form of $y = 1$ or 0 with 1 being the default class. However, when the transformation function is introduced, the regression is termed as logistic function $h(x) = 1/ (1 + ex)$ forming an S-shaped curve on a graph. Logistics regression algorithm is more so used in unpredictable data sets such as learning association like whether students are going to pass or fail in an examination.

Decision Tree

Decision tree also referred to as Classification and Regression Tree (CART), use nodes which represent input and output variables. The model is quite unique and uses multiple nodes; for example, nonterminal nodes represent input variables while leaf variables are for outputs. The decision tree is essential in creating probabilities of the future with the form. For example, one can use the tree to determine if you can buy a car in the event of another need which is also crucial. In this case, decision trees are beneficial to the prediction of future events, especially in the mining sector.

Support-Vector Machine (SVM)

This is another form of supervised machine learning algorithm of data analysis essential for classification and regression analysis. SVM assigns more variables to a given context with similar or related data, therefore, making it non-probabilistic binary linear classifier. The categories are separated by a clear and reasonable gap for ease of identification of differences. SVM are conventional solvers of modern problems and widely used in biological and related sciences, image classification, text and hypertext classification and recognition of handwriting patterns.

Naïve Bayes

The Naïve Bayes provides an algorithm where you can readily calculate the occurrence of an event when another has already happened. This supervised learning uses the Bayes theorem, which uses dependent and independent variables which determine the existence of the event. Naïve Bayes is mostly used as a weather forecasting tools as it predicts the coming events based on the previous happenings of the weather.

K-Nearest Neighbor Algorithm (KNN)

KNN is another machine learning algorithm which uses an entire dataset as a single training set and test set for a given computer. When retrieving the data needed, KNN uses dataset as a whole to find the K-nearest variable to the new event or those similar to the variable. The outcome or reward is usually averaged, and the results portray a mean or model which solves the problem at hand. KNN is generally applied in situations where you are searching for similar files such as in semantical documents and recommender systems.

K-Means Algorithms

This is unsupervised learning which groups similar data into clusters and calculates k value of centroids which gives the clusters a considerable distance between clusters. Initially, you will have to choose the value of k then assign each data point and compute to obtain clusters matching the value assigned k. Assign points between cluster centroids on each group and calculate the correct distance between them. Besides, ensure there exists not switching of points. K-means are most suited during clustering of big datasets and successfully used in astronomy, market segmentation, and vector quantization.

Random Forest

Random forest entails several decision trees represented by a multitude of nodes which suggests input and output variables. Random forest algorithms are created when multiple models are connected to develop a massive connection of trees of datasets, therefore, referred to as a forest. The trees are initially sampled to obtain the similarity and then connected with through nodes. Some of the applications of random decision forest algorithm include diagnosis of faults in machines such as engines as well as in the health industry to determine problems in diabetic retinopathy.

The Buildings Blocks of Machine Learning Algorithms

Machine learning uses applied mathematical basics, especially in the development of algorithms used in both supervised and unsupervised learning as well as reinforcement learning. More so, these algorithms use scalars vectors, matrices, and tensor to create variables which are interconnected in different forms and bring out the machine learning models. The general term algorithm signifies several mathematics applications derived from different areas.

Tensor, for instance, is a topic in mathematics which primarily deals with an algebraic object similar to vectors but double spaced and may take multiple forms. Another primary component of machine algorithm is scalar, or scalar quantity both used in mathematics and physics derived from single elements and has magnitude. Vectors are the most essential applied in multiple areas with both magnitude and direction and may include algebra and connected to other algebraic expressions or objects. Lastly are matrices comprising of rectangular arrays, expressions, numbers or symbols organized in rows and columns. The four are considered the primary building blocks of machine learning algorithms giving machines the ability to simulate human behaviors effectively.

Deep Learning and Cognitive Computing

As mentioned, deep learning is the broader machine learning method and comprises of artificial neural learning and machine learning on both supervised and unsupervised machine learning algorithms. Deep learning, therefore, consists of deep, recurrent, and convolutional neural networks. On the other hand, cognitive computing is where the machine applies the knowledge gained from cognitive science to create the architecture of several artificial Intelligence subsystems, including machine learning and human-computer interaction.

Like machine learning, both deep learning and cognitive computing use algorithms which alter the functions of how programming and coding functions giving the machine a new form of how to perform without human interaction. When correct datasets of variables have been fed into the computer, it readily incorporates the information and begins to learn on how to act on such events and develop inclusions without interaction to human. Therefore, several machine learning methods have become successful with machines now being able to respond and create things without inputting specific dataset parameters.

Basic Statistics and Probability Theory

Statistics and probability may have broader differences between them in mathematics, but when it comes to machine learning, they have several similarities. However, they do have dissimilarities depending on the algorithms used in each model. Probability is a fundamental aspect of machine learning and has been used to refer to the possibility of occurrences between happenings of two events. The prospect has always been termed to have values between 0 and 1. However, machine learning changes this narrative to ascertain that we can actually understand the outcome of a possibility when we jaw the right measurements.

When it comes to statistical data, a similar aspect can be adopted, but machine learning also depends on statistics which enables the process of classifying similar data more so on supervised learning algorithms. The theory behind statistics and probability of machine learning suggests that machines not only learn and remember the same event in the future but also utilize the probabilistic approach to determine if the event can occur again. The statistical theory indicates that machined considerably group data in the form which build an organized type of data for quick retrieval.

Deep Learning for Intelligent Agents

Deep learning, as stated, is the most comprehensive form of machine simulation and the most effective when it comes to machine learning. Since the introduction of deep learning, it has been used in the most complex and extensive areas with the benefit of effectiveness. When mentioning about deep learning intelligent agents, deep learning works as a machine which gather information regularly or when instructed by the user in real-time. An intelligent agent is therefore described as a program, usually a bot, which can make decisions or take actions depending on its surroundings, experience, or as instructed.

For example, games are developed daily with algorithms which utilize deep learning to either control or enhance the play. However, deep learning is used to act as intelligent agents to create more benefits to the game, especially for the game to respond to human interaction. When games become more interactive, so does the play become more interesting. In this case, the intelligent agent built within the game can make decisions and perform tasks or moves with the aim of defeating the player. Another form of an intelligent agent is used in the finance sector, where they readily react to the market value. That is, when selling your property or trading, prices or points may either shoot or drop henceforth affecting your indicated value. Therefore, intelligent agents play a significant role in moving with the trend where it readily modifies your value depending on the market changes.

Technical Requirements of Deep Learning

Keras

Keras is an in build Python library created by Francois Chollet with the aim of facilitating immediate experimentations. The library can run along with TensorFlow and Theano, among others and supports several neural networks, including convolution, recurrent, and dense layers. Keras is essential in machine learning, as it enables different methods, including translation, speech, image, and face recognition. Some of the benefits include fast and straightforward prototypes, built-in support for multiple GPU training and has fully configurable modules, among others.

TensorFlow

This is the most common open-source libraries created by Google and essential for numerical computation. TensorFlow is written in C++ and Python and excellent in sophisticated events, for instance, during the creation of numerous neural networks more so on random decision trees. The algorithms used in the library provide room for voice and image recognition as well as text-to-speech applications. Some of the benefits of TensorFlow are multiple documentations and guidelines, support of distributed training and model serving and recommended by a large community of developers and tech firms.

PyTorch

PyTorch replaced the Torch library of Python and is written in Lua competing significantly with TensorFlow. This library was created by Facebook and used in top facilities such as the University of Oxford and Salesforce. PyTorch is mainly used for training deep learning algorithms and used may accompany a more significant percentage of researchers. Some of the benefits of PyTorch are modeling process quite direct and transparent, supports distributed training, model micro-services, direct embedding, and portable tools for development enhancements.

Examples of Deep Learning

Automotive Researchers

As mentioned, deep learning has gained popularity in the automotive sector, especially with the ability for engines to self-drive without human interaction. Deep learning in vehicles, trains, ships, and airplanes have been on the increase without human help control. You only sit and use your voice for the engine to start and move. These automotive are fitted with sensors, navigation, and other parameters which ascertain their movements without causing accidents or other related harm.

Medical Research

There have been numerous studies conducted to prevent, cure, and eliminate the dangers of cancer which are on the rise. As such, cancer researchers are utilizing deep learning to automatically detect cancer cells more so when they start building up for easy identification and elimination. In one instance, the UCLA team created a modern microscope which provides high-dimensional data, therefore, providing training of deep learning detect cancer cells.

Aerospace and Defense

Deep space has also been introduced in aerospace and defense, providing multiple benefits to these areas. For instance, deep learning enables astronauts and related researchers such as at NASA to detect objects caught by satellites readily. In defense, deep learning has played a significant role in mapping areas where troops are to visit and determine if they are safe or unsafe, therefore preventing the loss of soldiers in military missions.

CHAPTER 3: DATA ANALYTICS

"I stay away from the arts... writing songs, being creative - those are downloads from God. You can't do data analytics on art"
Troy Carter

In this chapter, you will learn about the concepts of data analytics and its application in Artificial Intelligence (AI) and Machine Learning (ML). You will also learn about the different types of Data Analytics as well as the differences between a data analyst and a business analyst. Read on to find out!

Definition of Data Analytics

Data Analytics (DA) refers to the process of examining sets of data to determine the information they contain with the help of highly specialized software and systems. Data analytics techniques are popularly applied in commercial industries to enable companies to make more informed decisions. Scientists and/or researchers also apply it to approve or disapprove theories, scientific models, and hypotheses.

Data analytics, as a term, refers to an assortment of applications, from reporting and online analytical processing (OLAP), basic business intelligence (BI) to different types of advanced analytics. The expensive use of the term, however, isn't universal although in some cases, the term data analytics is used by people to specifically mean advanced analytics treating Business Intelligence (BI) as a different entity.

Having known what data analytics is, let's look at the various types of data analytics.

Types of Data Analytics

Raw data can easily be compared to crude oil. Nowadays, anyone or any institution with a reasonable budget can collect huge volumes of raw data, but the collection should not be the end goal. Companies and institutions that can be able to get extra meanings from the raw data collected are the ones that can compete in the modern day's complex business environment. At the heart of data collection and seats a very important phenomenon: data analytics. It is what makes the entire process have importance.

There are various types of data analytics, as explained below, read on to find out!

1. Descriptive Analytics

The major focus of descriptive analytics is summarizing what happened in an organization or institution. It examines the raw data to answer different questions like what is happening? What happened? and what will happen?

Data analytics is characterized by business intelligence and visuals like pie charts, bar charts, line graphs, or generated narratives. An illustration of data analytics can be determining credit risk in a bank or supermarket. In such a scenario, previous financial performances can be carried out to foretell the client's financial performance.

Data analytics help provide insights into the sales cycle, like categorizing customers based on their history and preferences.

2.Diagnostic Analytics

Just as its name suggests, diagnostic analytics is applied in determining why something happened. For instance, when conducting a social media marketing campaign, you might want to access the number of reviews, followers, likes, and mentions after the campaign. Diagnostic analytics helps you filter numerous mentions into a single view; it breaks down the entire information to simple bits that you can comprehend within a few seconds.

3.Prescriptive Analytics

As other data analytics gives you general insights on a certain subject, prescriptive analytics provides you with laser-like focus in answering questions. For example, in the health industry, prescriptive analytics can be used to manage the patient population by determining the number of patients who are obese clinically.

It allows you to add filters in obesity like diabetes and cholesterol levels in order to find out areas the treatment should be focused.

4.Exploratory Analytics

This is an analytical approach that primarily focuses on establishing general patterns in raw data so as to identify features and outlines that might not have been discovered using other analytical methods. To use this approach, you must understand where outliers are occurring and the way different variables are related to making well-informed decisions.

For instance, in biological monitoring of data, websites might be affected by different stressors. Therefore, stressor correlations are very important when you try to relate the biological response variables and stressor variables. Scatterplots and correlation coefficients provide you comprehensive information on the relationship between the involved variables.

When analyzing different variables, however, the basic ways of multivariate visualization need to provide deeper insights.

5. Predictive Analytics

This is the use of machine learning techniques, data, and statistical algorithms to determine the probability of future results basing on historical data. The primary agenda of predictive analytics is helping you go beyond what as happened and give the most logic assessment of what is more likely going to happen in the future.

It uses recognizable results to come up with a model that can predict figures of different types of data as well as new data. Using recognizable results is important because it gives predictions that represent the likelihood of the targeted variable based on the estimated significance from the provided set of variables.

Predictive analytics can be applied in the banking systems to establish fraud cases, maximize the cross-sell and up-sell opportunities in a company, and measuring levels of credit risks.

This will help retain valuable clients to your business.

6. Mechanistic Analytics

Just as its name suggests, mechanistic analytics enable big data researchers to comprehend and understand clear alterations in procedures that can result in changing of variables. Equations in physical sciences and engineering determine the outcome of mechanistic analytics. They also allow data scientists to determine the parameters if they understand the equation.

7. Causal Analytics

They allow big data researchers to figure out what can happen if a single component of the variable is changed. When you opt for this approach, you should rely on the number of variables to determine what is going to happen next. The approach is appropriate if you're dealing with extremely large volumes of data.

8. Inferential Analytics

Inferential analytics takes different theories into account to determine certain aspects of the large population. When you use this approach, you will need to take a smaller sample of information from the target population and use it as a basis to infer parameters about the rest of the population.

After looking at the various types of data analytics, it is time you learned how data analytics is applied in different aspects. Read on to find out!

Data Analytics Application

Data analytics methodologies include The Exploratory Data Analysis (EDA) that aims to establish patterns in data and the Confirmatory Analysis (CDA) which uses statistical techniques in determining whether hypotheses about a certain set of data are false or true. While EDA is usually compared to detective work, CDA is similar to the task of a judge during a court trial.

Data analytics can be separated into Qualitative data analysis and Quantitative data analysis. Qualitative data analysis focuses on understanding the content of non-numerical data like images and audios. Quantitative data analysis, on the other hand, involves analyzing numerical data with quantifiable variables that can be measured statistically.

Data analytics can be applied in providing business executives and corporate workers with information about business operations, key performance indicators, customers, among others. This is done through BI and reporting. Previously, data reports and queries were mainly created for end users by BI developers, but today, organizations are increasingly using self-service BI tools let executives, business analysts as well as operational workers run their quarries and come up with reports by themselves.

Advanced types of data analytics include data mining: this involves going through large sets of data to identify patterns, trends, and relationships: Predictive analytics that seeks to predict customer behavior and machine learning: an AI technique that automated algorithms to sort through data set quicker than it can be done through conventional analytical modeling. Therefore, big data scientists apply data mining, predictive analytics, as well as machine learning to sets of big data that contain structured and semi-structured data. Text mining also provides a means of analyzing emails, documents, and other text-based content.

Data analytics have a wide range of business use. For instance, credit card companies and banks can analyze withdrawal and expenditure patterns to detect theft and prevent fraud.

Commercial companies and other marketing services providers do clickstream analysis so as to identify website viewers that are more likely to buy a product or service based on navigation and page-viewing sequence.

In addition, mobile network operators can examine customer data to forecast churn in order to prevent defections to rivals and boost customer relations. They also engage other companies in analyzing CRM analytics in order to fragment customers for marketing campaigns and equip their call centers with up-to-date information about callers. In healthcare, data analytics has also been used by organizations to mine patient data in order to evaluate the effectiveness of treatments to diseases such as cancer, AIDS, among others.

What Role Does Data Analytics Play in AI And ML?

Artificial Intelligence refers to the general field of algorithms in which machine learning is the leading inventory at the moment. AI is just a computer that has the ability to mimic or simulate human behavior or thought. Within that, there is a subset known as Machine Learning (ML) which is the most exciting part of Artificial Intelligence.

So how do ML, AI, and data analytics intercept?

ML is a branch of AI where a class of data-driven algorithms enables software applications to produce highly accurate data without any need of explicit programming.

The basic factor here is to build algorithms that can receive input data and induce statistical models to predict an outcome.

Data analytics employs science disciplines like mathematics and incorporates techniques like cluster analysis, data mining, and machine learning.

So, machine learning is a subset of AI, while data analytics is an interdisciplinary field used to extract meaning from data.

What Is the Difference Between A Data Analyst and A Business Analyst?

People often confuse between data analysts and business analysts. In some occasions, the two words are used to mean the same thing which shouldn't be the case. To find out the difference, let's look at each one of them and find out what it means.

Business Analyst

They do a lot of things in the business world. Their role primarily centers around requirement analysis. Some of the tasks business analysts do are; understanding genuine issues of stakeholders or business users, elicit and document business requirement, ensure completeness of business requirements, ensure that the technology team has comprehended the functionalities and is headed in the right direction.

Data Analyst

A data analyst manages to analyze large volumes of data, modeling data with the intention of coming up more useful business insights. The data analyst comes up with behavioral patterns and hypothesis on the primary basis of predictive analytics to support decision making.

They frequently utilize factual statistical models to make decisions and hypothesis.For instance, data analysis engages a lot of consumer-driven franchise and propelling their strategic business decisions. Subsequently, the outcome of data analytics forms the input of business decisions and essentially, how the system should be designed to suit those decisions. The driving factors of customer behavior or pattern, the designing systems these patterns are based on is governed by the efforts of a data analyst.

Although business analysts define the functionalities of the system, they solve issues by initiating the right feature in the system.

Skills wise, both data analysis and business analysis have contrasts and some comparability. Some of the skills needed for both roles are analytical mind, creating opportunities, the ability to see the master plan, a good business acumen, and presentation aptitude.

Both roles have an extraordinary future and bring a unique set of value to the table. A large number of organizations use both terms reciprocally and see them as the same. As seen from the above tasks and skills, they are quite different.

With more consumer data being accessible, data analysts have the upper hand in influencing the lives of people and make a difference to businesses and analytical business insights.

So, the answer to whether data analysts and business are the same no, you've seen the difference!

Qualitative and Quantitative Techniques

In data analytics, qualitative and quantitative analysis are two fundamental ways of data collection and interpretation. These methods can be used concurrently since they both have similar objectives. They might have some errors. So using them at the same time can make up for the errors each method has, leading to quality results.

There are overlaps in qualitative and quantitative analysis. I have outlined the similarities and differences between these two analyses techniques used in data analytics.

What is Quantitative Analysis?

This research analysis method is often associated with the numerical analysis where data is assembled, classified, and recorded for certain findings using a set of different methods. In this technique, data is chosen randomly in large samples and then comprehensively analyzed. The advantage when using quantitative analysis is that findings can be used and applied in a general population using the research patterns identified and developed in the sample. This is a disadvantage of qualitative data analysis because of limited generalizations of findings.

Quantitative analysis is objective. It aims to comprehend the occurrence of events and describe them using analytical methods. More clarity can, however, be obtained by concurrently using quantitative and qualitative techniques.

Quantitative analysis is mainly concerned measurable entities like length, width, temperature, weight, speed, and many more. Data can be represented in tabulated form, pie chart, line graph, or any other diagrammatic representation. Quantitative data can be categorized as either continuous or discrete, and it's often obtained using surveys, interviews, experiments, or observations.

There are limitations in quantitative analysis. For example, it can be difficult to discover new concepts using quantitative analysis and is where you need to apply qualitative analysis technique.

What is Qualitative Analysis?

It is concerned with analyzing data that can't be quantified. Qualitative analysis is about understanding the concepts of properties and attributes objects or participants. It gets a deeper understanding of "why" a certain phenomenon takes place. This analysis technique can be applied concurrently with quantitative analysis. Qualitative data can be wide-ranged and multi-faced, unlike quantitative analysis that is restricted by certain rules or numbers. When applying qualitative analysis, you must be well-rounded which any physical properties the study is based on. The typical data analyzed qualitatively include gender, color, nationality, appearance, taste, among others, as long as they can't be computed. Such kind of data can be obtained through interviews.

Just like quantitative analysis, qualitative analysis has limitations. For example, it can't be used to generalize the entire population. Only small samples are applied in unstructured approach, and they do not represent the general population; hence, this method cannot be applicable in generalizing the entire population.

The qualitative data analysis technique is anchored on the classification of participants according to attributes and properties, while quantitative analysis is based on the classification of data anchored on computable values. Quantitative analysis is objective while qualitative analysis is subjective.

Therefore, there is a clear cutline between the two. They are both applied in data analytics, making it more accurate and credible.

Data analytics plays a significant role in today's world. You have learned of the different types of data analytics, how it is applied in various fields like AI, ML, and banking. Towards the end, you found out the difference between a data analyst and business analyst, the roles each one plays and features that distinguish them.

CHAPTER 4: NEURAL NETWORK

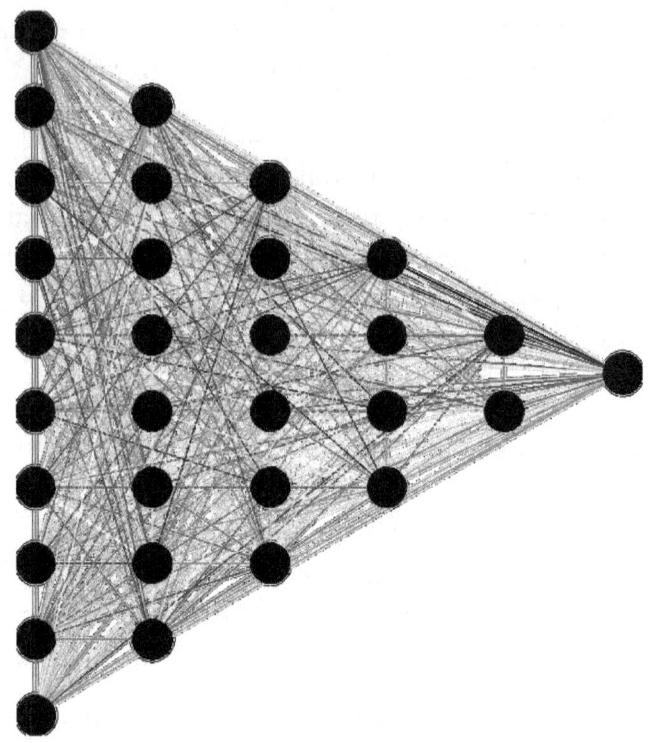

Neural Network Image

"The pooling operation used in convolutional neural networks is a big mistake, and the fact that it works so well is a disaster"
Geoffrey Hinton

A neural network is in simple terms described as either a software or hardware system programed to mimic the neurons found in human brains. The idea of neural network originated from two Chicago university researchers, namely Walter Pitts and Warren McCullough. It's several processors brought together to solve underlying issues in a series of trial and error. The idea of the neural network was majorly researched on different fields such as in technology and neuroscience sector in the early 1960s. The industry has recovered from its decline and is enjoying massive support in the technological world due to the modern era graphic chips. The chips have very high processing power disposal.

The neural network technology is applicable in different sectors fueling its resurgence. It is used in the analysis of a survey of oil, climate forecasting, and data analysis.

Neural Network Mathematics

Introduction

The idea of calculations involved in the neural network is not hard to work out; it all depends on the will, determination, and passion one employees. Most of the math in neuron network comprises of matrices.

1) Weights

Weights are the determination of neurons that transmit value. The size of the weight and value are directly proportional to each other. When the value being sent is low, then the weight is likely to be small and vice versa. The weight arranged in a matrix layout for calculation purposes. The neuron is usually placed in the input side.

For instance, an input layer which contains four neurons, and the subsequent layer contains five neurons. From this layout, a matrix can be formed of 4 rows and five columns. The weight values are then placed in the created matrix. The matrix can be named *P1* for easy identification. When there are different layers, then other matrices are formed with the same outline.

When a layer **P** contains **Y** neurons, the following layer **P+1** contain **Z** neurons then to calculate weight matrix will be **Y** by **Z** matrix.

2) **Bias**

In the decision-making process, individuals think of different factors before making the final decision. The outcomes of the decision settled upon are also mitigated. With all this thinking, some results are never expected. This is where Bias comes to use in neural networks. Neurons that are not within the input layer contain Bias. Bias and weight are a bit similar in that they both carry value. Bias can be put in matrix formats in the column sector.

Calculations

Let the layout be called Z. The formula abbreviations are as follows

- b – the value of Bias

- x - input to the neuron

- w – the value of weights

- n – The value of inputs from next layers

- i - value between o to the incoming layers

Typically, neurons that have value during the start of calculations are the input neurons found within the input layer. To obtain the other layers neuron values one needs to;

a) Multiplying all the inbound neurons with its weight.

b) Taking the values and adding them together.

c) Then adding the bias matrix of the neuron needed.

The above calculation involves obtaining Z. This method is hectic when it consists in calculating numerous neurons. The most straightforward tactic employed to carry out the challenging part involves;

The weight matrices formed can be used.

1. A weight matrix is formed that transmits value between the input layer and the output layer; e.g. (Y) by (Z) matrix.

2. A bias matrix is also formed (Y) by 1.

3. The input layer matrix (Z) by 1 is created.

4. You are transposing the weight matrix to form (Y) by (Z) matrix.

5. The dot product is determined. This can be done by obtaining the value of transpose of the input layer and weight. The outcome of (Y) BY (Z) matrix and (Z) by 1 matrix is (Y) by 1 matrix.

6. Add the dot product to the bias matrix. The value usually comes of equal measure if the calculation is done correctly.

7. In the end, one obtains the value of the neurons (Y) by 1 matrix.

The steps above should be repeated for each weight matrix and the neuron values until the last bit to obtain the output layer.

The explained method above can only be used in a connected neural network.

Neural Networks and Learning Machines

The Neural networks are a series of algorithms programmed to solve complex computations. The idea is to make it mimic human being neurons. The artificial neural network can be used in enabling machines to learn and recognizing patterns. The system derived from this idea works by analyzing value layer inputs.

The neural net can be presented in a graph format. It's made of nodes. Arcs link the nodes. The weight brings together the nodes and the arcs. The weight within the neurons transmits the values. The network of neurons forms the machine learning program. The signals being transmitted within the neurons usually are in chemical and electrical setups. The sector can help in regression of constant target traits. A neural network can be used in different industries such as in business, forecasting, and also obtaining data.

The neural network is made up of three layers. Numerous nodes make up a layer. An activation function is carried out by the nodes.

V Input layer – this layer involves the raw data collected and inserted into the grid. The input layer obtains value containing instructive units. The total sum of nodes found in an input layer usually tallies with informative units. Signals and information transmitted within the nodes do not encounter any distortion in the process. Duplication of values received then happens as it received only one signal. The duplicated data is transferred to its numerous output units.

V Hidden layer – it involves the process involved in input value and the weight value obtained in the unknown entities. The hidden layer can be more than one. The weights usually increase values within the hidden layer.

V Output layer – this layer can be determined by the events ongoing on the weights and the hidden layer units and also the output entities. Output layer connects to the hidden layer. When a value reaches this layer, it's returned with a value matching its forecast unit.

Types of Neural Network

There are various types of neural networks. Models that are easy are ones made their units are connected by two layers, namely output, and input. The output and input layers' units correspond to each other.

a) Feedforward Neural Network

This type of artificial neural network is the simplest in all the different types. The data in this layer transmitted singly. Data being transmitted goes through input nodes and leaves through the output nodes. Some Feedforward Neural networks contain in them hidden layers, while others don't. The propagation in this network forward-facing.

The calculations involved the addition of value inputs and weighted value to the output layer. Reflection of the output is only done if it has surpassed a definite amount which enables neurons to be activated or not to be activated in the process.

The Feedforward neural networks can be applied in the fields of visualization in processors and communication recognition in cases whereby the act found to be complicated. Its advantage is that it has low conservation level.

b) Radial basis function Neural Network

The Radial basis functions neural network is programmed to contemplate on issues involving the space of a plug. The center of the said plug is also factored. It's made up of two layers.

It uses the Euclidean model to calculate the distances involved. The radius is obtained to help in categorizing different levels. Where the point is not within the radius, then chances of a new position being categorized is low.

Its application involves being used in energy restoration grid. The energy sector has grown, and with this comes, complications such as electric blackouts. With many people and industry depending on power, restoration needs to take place fast.

The process of blackout restoration involves the following;

- It's usually the hospitals, institutions, police stations and also essential infrastructure such as train services that are considered first. Before any other clients, when power is being restored.

- The next step involves power stations that supply power to a large area.

- After essential clients, then power can be restored in households and estates.

Recurrent Neural Network

The response of an event is forecasted by the neural network saving the output layer then transmitting this data to the input layer.

With each step passed the neuron can remember a bit of data along the way, making it resemble a memory cell. The data collected is stored for some time to be used later. When the forecast is not right, two methods are applied to make the corrections, namely; error correction and learning rate. With these methods, the neural network makes amends to where it was wrong and begin to predict the correct assessment.

Recurrent Neural Networks can be applied in different sectors. For instance, in a text to speech adaptation. The deep voice was created through this neural network.

d) Convolutional Neural Network

Convolutional neural networks have parallel features to feed-forward neural networks. The Bias and weighted value in the neurons are directly proportional. The neural network is typically used image and motion processing.

The input variables are transmitted in the form of groups. The categorization process configured such that the neural network can recollect the images sent and more accessible computation. The operations being carried out comprises of HSI scale to Grayscale or RGB to Grayscale. The process involved help the input features are taken in batch-wise like a filter. This will help the network to remember the images in parts and can compute the operations. These computations involve the conversion of the image pixel format from RGB or HSI scale to Gray-scale. The format transformations make the program to be able to detect images. Classification can after that happen.

e) Modular Neural Network

This artificial neural network is different from the other systems, and this is because the modular neural net is made of various networks that are operating unconventionally yet adding to the output of the network. The various neural networks found within the modular system have different characteristics making the input. The subnetworks are not submitting electric or chemical signals to each other nodes while carrying out its functions. The modular neural network is preferred to other networks due to its ability to deconstruct immense computational. These allow for the complex task to be handled with ease. The process carried out improves the quality of computation by ensuring the nodes transmit faster. Amount of neurons involved will also determine the speed and quality. This kind of network has seen an increasing level of research done to be able to understand it more.

Steps one can take to build a Neural Network

1) Collecting data for training

One needs to receive a lot of data. Below are ways the neural network can be able to adapt to the pattern. The values are named A and B.

a = np. array([-2.0, 1.0, 2.0, 3.0, 4.0, 5.0])

b = np. array([-4.0, -2.0, 1.0, 4.0, 6.0, 8.0])

2) Construct the neural network

To be able to build the neural network model one needs to choose on the number of neurons and layers they want to build.

- model = tf.keras.Sequential()

- model = add (tf.keras.layers.Dense (units=1, inputs_shape= (1)

- model.summary

3) Model compilation

In this step, one chooses the type of algorithms needed, such as optimizer. The neural network will also require the loss function.

Model.compile (optimizer= 'sgd', loss= 'mean_squared_error')

4) Enter the training data collected

The data that is fed into the model will enable the neural network to search for a corresponding function. The number of connections to be made by the built neural network is to be decided.

Model.fit(a, b, epochs= 500)

5) Test the model

The built neural model, when completed, needs a test. This can be done by exposing the model to raw data. For the neural model to the considered a success, it needs to speculate the new data output precisely.

Human Brain Vs. Neural Networks

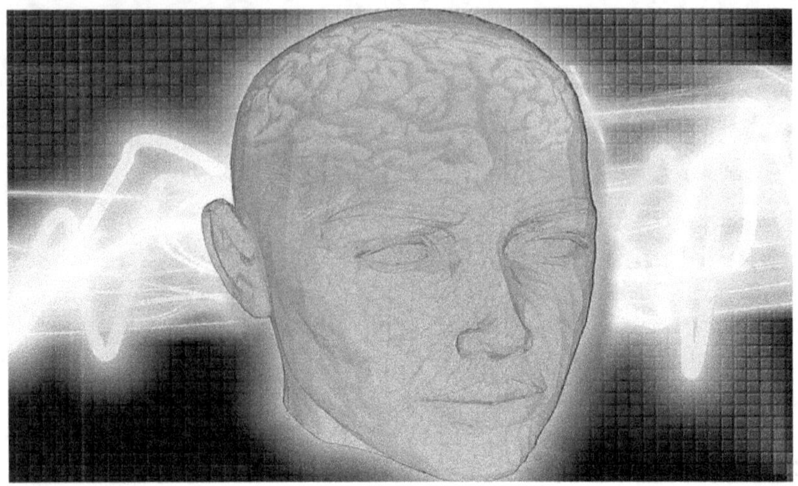

Level of Creativity

Creativity is usually explained as the ability of someone or something to develop concepts that are new or coming up with systematic models that constructive.

In the modern era, people are findings ways to develop technologies that are considered creative. The computers itself applying creativity in industries. With this, computers have been able to challenge some scientific theories, creating music beats and also developing approaches by are useful.

Computers though are not capable of knowing what the best new idea is. Modern computers haven't produced any new scientific theories, only challenging some existing ones. The human brain still has an advantage over it in this sector.

Ability to have Emotion and empathy

These two feelings of Emotion and empathy are yet to be fully comprehended by the neural networks. Emotion is an essential aspect of being able to forecast what events can happen, and it also acts as a form of controller. The level of accuracy to detect facial gestures by the neural networks has increased immensely. In the modern world, some technology companies are trying to develop robots that can provide company to people.

Process of creating a plan and executing it function

The function of Planning and executive ideas is an essential part of the human brain. This function originates from the pre-frontal lobe. The human brain frontal lobe came into existence and is found in Homo Sapiens Sapiens. The ability of the human brain to plan and execute cases of people being able to speculate on the future event and outcomes.

Human brains are less likely to make a rational decision compared to computers. While this is the case, the ability of humans to process emotions better than machines enable them to be way adaptable and better in making plans and being able to execute them. Computers are limited in carrying out this function as the speed level is not considered.

Consciousness

Consciousness is defined as the self-referential involvement of our reality and choices we make and able to feel at each particular time. It's what gives people motivation in the form of spirituality and level of surprise. Different physiology fields have tried to research and get to more about consciousness, but challenges are faced. Questions such as what it is. How did it come by? Where did it originate? What requirements are needed to reproduce it? Are asked continuously but without success.

With all these questions unanswered, the likelihood of consciousness being created in computers hasn't begun. Some scientist is of the idea of starting with cracking the emotion and execution function. Ethical challenges.

Hardware or software-based Neural Network

A lot of work has been carried out to be able to make artificial neural networks surrounding replications on the technologies. When one outlines sequential ANN simulators obtained from machines. The Intel Pentium processor is one of the neural networks which is growing exponentially due to its high-performance standard. Its form of conventional von-Neuman processor. The fastest processor though is not likely to bring about real-time reaction and the learning of the network which have a large number of neurons. The opposite happens for parallel processors with many modest processing elements.

Applications of hardware neural networks involve communication recognitions in items and also devices. The hardware neural networks can also operate a large number of orders at a faster rate compared to neural software networks.

Advantages of Artificial Neural Network

- ❖ Storage facility for the system- data is stored in the system such as modes of programming in the whole network. When some data goes missing, it doesn't affect the function of the net.

- ❖ The network can function with limited information-
  the moment the neural network has been feed with
  training data; a successful model will produce outputs
  even with less information. The only limiting factor is
  affecting the performance hinges of critical data
  inserted.

- ❖ The Artificial Neural networks are straightforward to
  apply.

- ❖ Level of tolerance for errors is high- when a particular
  cell in the artificial neural network gets distorted, the
  data output gets produced still.

- ❖ Memory function- the ANN can learn. This can be
  achieved by training the ANN by showing illustrations
  to the system. The proportionality of the network to the
  shown illustrations is direct.

- ❖ Slow distortion- the ANN gets corrupted with time
  although this happens slowly.

- ❖ Ability to train and Learn- the ANN can learn from
  examples shown and from that training able to make
  better choices when faced with a similar scenario.

- ❖ Parallel processing ability- the number of networks in
  a system can enable it to execute different functions.

The Disadvantages of Artificial Neural Networks

- ❖ Level of dependence on Hardware- the artificial neural
  networks depend much on processors to execute its
  functions properly. The processors need to behave a
  parallel processing power.

- ❖ Unsolved functioning of the system- the artificial neural network is not capable of explaining how and why signs when a solution is searched. This limit is the most critical in the network system.

- ❖ The level of assurance for the structure of the network to be outstanding- there are no policies put in place to regulate the kind of artificial neural network structure required. The ideal stricter is usually obtained through trial and error assessments.

- ❖ Reinstructing Artificial Neural networks are robust- Adding data at a later time will likely bring corrupted results.

- ❖ The challenges faced to give an illustration of problems to the neural network model- the most challenging part is converting problems into numerical values before feeding to the artificial neural network. The ANN only absorb data in a binary format. The accuracy of the model working well in this case all depends on the ability of the user to convert useful data.

- ❖ The period of network function is relatively unknown- the ANN system sometimes comes up with error as a value which is a limit.

CHAPTER 5:
PREDICTIVE MODELLING

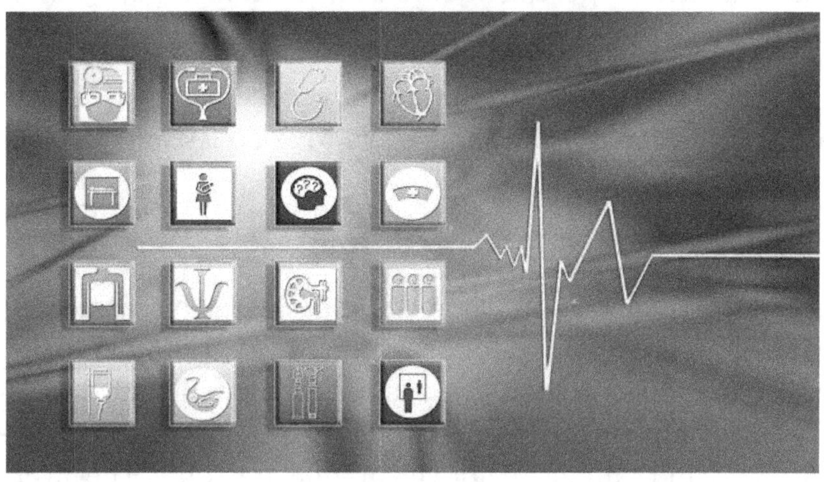

"Science is a little bit more than a wonderful way of modelling and predicting; it's a wonderful technical abstraction. I think science is a really wonderful technical abstraction"

Bernard Beckett

Predictive modelling is a tool used in Predictive analytics in a process that includes already collected/tabulated data from the past or current to draw future outcomes. It involves a data mining process that tends to forecast the possibility of getting new results or remain the same in the future.

Once data is collected, a statistical model is created to help make the predictions/ guesses on a future trend or behaviour related to certain data at hand. For example, growth in technology. With the recent trend of social media use, they can use the numbers or data from phones, web searches and so on to predict the data and come up with an assumption of what will happen for like the next ten years. This helps in deriving changes or solutions to solve the issue or improve it.

So as to come up with the right value or results for the future, the statistical model can be revised from time to time, adding results from other sources to better its performance.

Predictive modelling is accurate in its predictions as it gives accurate forecasts by allowing users to ask questions or make assumptions.

Predictive versus Interpretation

Predictive is about anticipating, estimating, or forecasting what may come about in the future. On the other hand, interpretation is asking a series of questions about data. It involves implementing processes in which data will be reviewed and sampled before its original outcome. They work together to ensure e data being used in the model is of the highest quality.

Techniques of Predictive Modelling

Predictive modelling involves two main techniques which include:

Regression technique

It is mostly used in statistics to find the effect of one variable has on the outcome. Using the linear regression, they can use the data to determine the Variable on the relationships these variables have to one another. By using the normal distribution, it is able to find the specific factors from a large number that affects the product. On the other hand, logistic regression, the outcome is determined based on the results of other variables which are discrete to get a prediction. It mostly follows the data from the already existing list to determine the only limited response based on a categorical number. The number is always 0 or 1 or in multiple regression where the numbers can change to 2 or 3.

Neural Networks

This technique is used as the Artificial Intelligence (AI) model to handle complex functions. They can be applied in classification, control, and prediction in different sectors of neuroscience, finance, and medicine. In this case, it works without knowing the relationship between the input and output. There need to be training, both supervised and unsupervised.

It works on the basis of Artificial Intelligence. For instance: smart assistants, image recognition, and natural Language Generation. It's a powerful model with the ability to review a huge amount of labelled data in search of correlations between variables.

Application of Predictive Modelling

Predictive modelling can be applied in different sectors and guarantee the best results. It has been used before, and it has resulted in positive results. It can be used not only in the financial institutions but also in educational organizations and businesses.

It can be easily applied in the following areas:

- **Health**

Health sectors can use predictive modelling to determine the life expectancy of cancer patients which can help the physicians with treatment options. It can also keep the records of those patients with serious health issues and risk being admitted from time to time.

Healthcare providers can be able to predict the effectiveness of new medical procedures, medication, treatment options, and tests. This can also help them improve the health sector and predicting a healthier and improved health sector.

Moreover, hospitals can be able to monitor their patients' records or their employees. This will help them have all their data and predict their possibility of retaining their job in the next five years or monitoring the health of their patients with ease.

· **Insurance**

For the health insurers, they can be able to determine the health status of their customers and determine which of them is more at risk and needs special care. They can detect any danger with chronic illnesses as well as fraud in the health sector. Using the data of their clients, they can decide what best suits them in case they have an attack next time. They mostly rely on the medical; results in the system and get a probability of who will need a health provider to take care of them.

Insurance companies prefer it for retaining customers, determine the premium rates, and detect fraud. Insurance managers use it to grow and avoid risk by safeguarding the lives of their policyholders at any given time. For a vehicle or bike insurance, they will predict the chance of it serving you for long or any cases of accidents.

· **Customer retention**

Recently, there is a lot of competition, and ensuring your customer is satisfied and awarded should not be the case. You should not wait for the customer to complain about your products and terminate your agreement to act. You need to make sure predictive modeling as you will learn about the probability of the customer staying and what you can do as well as their behaviours.

- **Telecommunication**

They can use it to retain more customers. They can use the data on different media platforms to help create good relationships and new customers for their cross-sell/up-sell. In most cases, they concentrate on their most esteemed and valued customers.

- **Media and entertainment**

They can use it for determining trends, influencing attributes, and how the sports and entertainment sectors are fairing. They can get all the information about entertainment in different sectors easily than before, which could take an entire week to arrive at a finding. In most cases, media can negatively influence people if the content is not right for them. This greatly helps as they are able to determine appropriate audiences for a certain content without much effort.

- **Manufacturing**

For any manufacturer, you may need all the information on the production process, what is needed in terms of resources, manpower, and skills. You can predict the outcome of the product as well as determine their failures and the possible solutions to the problem. You can also use the information on making sells and targeting the right people for your products. It can guarantee you data on the quality and quantity of products you have done. That can help predict the number of products or assets you will be able to produce after a given period of time.

- **Utility companies**

It helps detect any malfunctions in the equipment used, future solutions to the problems, resources required, improve the safety of the equipment or resource and reliability as well as the risks involved. Understanding this data will predict the outcome of the problems and how to solve the issues before the case.

- **Retailer**

You can decide on matters of goods, the availability, and demand and how to grow your brand using campaigns and other marketing strategies to get customers while retaining those you already have.

- **Child protection**

The child welfare agency uses this tool to curb child abuse and neglect. In the past, there were cases of neglect and abuse, as the people involved did not face the law. Nowadays, welfare ensures that children are protected and in case they are abused, and they can report easily and get help.

- **Sports**

Sports, like any other area, has embraced the use of predictive modelling id predicting sports lineup and the possible winner. Using the information from their past games and the recent ones while comparing their strength and weaknesses could help analyse results for future tournaments or line-ups.

Predictive Modelling Process

This process involves a number of steps to make prediction easy and effective. They include the following:

1. Understand your business objectives

2. Define your modelling goals

3. Collect data

4. Prepare your data

5. Determine the required variable

6. Select and built models

7. Validate models

8. Use models

9. Monitor the performance

Predictive Modelling Tools

Predictive tools have, in recent times, become popular for giving data its meaning. Unlike in the past, where it was less used due to lack of enough skills and were restricted to IT. These tools are easier to use, and they ensure your prediction makes sense. Most business people and those from different sectors have established a relationship with these tools by developing new ones to help make prediction easier and more worthwhile. These tools help detect problems and give solutions on how well to handle the problems.

These tools are differently designed depending on your needs. There are those created for the experts, and others can be handled with anyone. These include:

- **TIBCO Software**

It helps the brand make the right decisions basing on the data of their customers. They mostly look at the trends and behaviour of their customers and know how well to retain them. What will fit them and if they will stick for long or not?

- **Terracotta In-Genius**

It helps organize data in an as easy and effective way. It groups the data in different categories to make prediction easy and fast.

- **Medalogix**

This tool is used in the hospitals to make the right decisions concerning patients stay in the hospital as well as their treatment using their records.

- **MyCityWay**

It helps brands and organizations set up their mobile app to use for referrals and campaign marketing.

- **Medio platform**

They help understand the reason as to why their customers have left their website by knowing their behaviour. Mostly, it works by making the brand understand what their customers are yearning for from them.

- **SAS Text Miner**

It helps brands arrange and categorize their data in a more quickly and effective way in regards to their requirements.

Data Preparation for predictive modelling

Data preparation is an essential part of predictive modelling as it tends to clean up and select the right data for your prediction. The data to be used here should be of high quality and meets the acceptable terms.

Since the data to be used should be clean, relevant, and on point, you may need to follow these steps to ensure you have the right data for your predictive modelling:

- **Identify the data**

Once you have identified the data you need, ensure it is the one you seek as data ends to be found in different formats and styles at different sources. A source will direct you to what you need.

- **Look for ways to get the data**

Probability is that the data you need is not yours. You need to look for ways to contact the owner of the data. If its data from a raw source, you can collect it and make the right changes to make it worthwhile.

- ### Decide on what variables to use in your analysis

As you may not know which set of variables will give you the right results, you can consider using a variety of variables to determine which one actually works for you.

- ### Decide whether to consider derived variables

Ensure you use the already known variable than that which has never been used before for better results.

- ### Know the data you collected

Knowing your data will help you determine your output. High quality and accurate data will help you get the right predictions. It will give you the clarity you need in choosing the right algorithm for your model. You should ensure your data is complete, has no errors and has no missing characters for the best service.

- ### Choose the right algorithm

Once your data has established the right relationship, it is now ready for predictions. You should note that information that does not have a relationship with the model; there will be challenges along the way.

Predictive Modelling Case Study

Predictive modelling is related t a lot more than the business world. Through its techniques, it can help come up with future solutions of an issue through prediction. Using a wide number of data from different sources, you are sure to get the best future outcome.

Here are some of the case studies of predictive modelling in:

Actuarial science

Predictive modelling related to predicted and explanatory variables. It mainly deals with financial events in the past and the present to come up with a prediction in insurance as well as other risk management applications. In matters of insurance, predictive modelling helps you understand the risks involved in the insurance sector as well as having an idea on how to save your customers from leaving.

When dealing with property and casualty insurance, you can be able to predict the cases of accidents and the casualty as well as the loss of property through fires, floods, and so on. From the prediction, you can plan for any accidents and ensure your policyholders are helped immediately they pass through property damage or accident. You will also learn of the risks involved in the process and learn t prevent it.

Drug sensitivity

Use of prediction in the treatment response of cancer patients is an important part of predictive modelling. Having their data on various treatments may help doctors make an easy decision on what kind of medication they should be given and what therapy would work for them.

It targets the types of alterations in cancer patients such as gene sequences and point mutations and their ideal therapies. Response to medication may help biologists and doctors to research on what compound is causing the reaction and how well to solve it. You will get to know of the side effects of the drug on different cell lines, humans or animals.

Drug sensitivity refers to the extreme response to treatment. Once you select the variables to be used and the research from different research, you can implement the model that will predict the outcome of the drug sensitivity either from clinical tests or research.

Marketing

Marketing can be used in predictive modelling to improve the results of the organization or business. In recent times, more business people have trusted Predictive modelling. They can use it to campaign, improve the operations in your business, and know how to retain your customers as well as getting new ones. Retaining your customers has been found to increase your profit and using the right data to predict your future could help you keep solutions in place for problems you may encounter on your way and be financially safe as you will be prepared for whatever may come.

It enables you to learn how to attract your customers and what to do to remain in a highly competitive world. You will learn to take risks to grow your business as well as give your customers what they need.

How to build a simple predictive model

A predictive model is an amazing way of coming up with the correct outcome. You use the data you have to predict the future outcome. For these results to be successful, there need to be factors that should influence the possibilities of your outcome to happen. As it involves machines to help in prediction, you need to sort data correctly to give you a perfect result. First, you have to build the model to work effectively. Here are the simple steps on how to go about it:

1. Data

For a successful predictive model, there needs to be data so as to enable the prediction. You need to collect data that can help make correct predictions. As much as the data used can be in bulk, you may have to table it well to enable the model to work on future outcomes.

2. Model

It will give us the prediction results we are looking for. It incorporates computer algorithms in coming up with the correct future outcome. It can handle a set of data to make predictions, and the moment it is built, you can be able to use to predict. The model will be used for learning and giving results.

3. Prediction

Prediction is the assumption that will be made. The date you have collected is what will be used to make predictions on the outcome you can use a different or same set of data to get results. Predictions will be made after the model algorithm is used to relate the data and make a future prediction.

Why implement Predictive Modelling?

Recently, more and more business people have embraced the use of Predictive modelling techniques to keep their growth. It is no longer a thing of the mathematicians or physicians as it has been proven effective due to their positive results in different organizations. As you need to give your computers your data to come up with a model to provide you with future outcomes; you will prepare well according to the results not to be affected negatively.

It comes with a number of benefits worth trying. For instance:

Reducing Risk

You will be able to monitor your credit scores and analyse your customers' credibility. You will be able to understand the losses you have recurred in the past and get to know how well to solve the issue in the future to avoid negative effects on your credit score.

Improving operations

Most companies may consider using predictive modeling to check their inventory as well as manage and save resources. Some use it to make its operations easy and easily accessible by their customers. For example, airlines use it to set the time for arrival and departure as well as the price. Hotels and restaurants can predict the number of guests they will receive in a day and their service; food and accommodation and table their income easily. It makes it easy for you to manage your inventory by analysing the profits and loss without looking into files.

Detect Fraud

Fraud may have a negative impact on your business. Predictive modelling allows you to detect any criminal behaviour in your business. With the increase in technology, cyberbullying is an order of the day. So as to prevent this, this process helps you spot any underlying threat and enables you to solve them on time.

Optimize Marketing Campaigns

For any business to grow, you need to invest in your marketing strategies. Predictive modelling will help you grow your business by attracting potential customers and retaining regular customers. They help determine customer responses on your products and their possible purchase.

CHAPTER 6: CYBER AND NETWORK SECURITY

Computer Network refers to the interconnected group of computers and other devices connected to share information. However, there are different types of network depending on the location of individual computers and devices. Firstly, there is Local Area Network (LAN), which involves a smaller geographical area like an institution. Secondly, there is Metropolitan Area Network (MAN) which involves a relatively larger area like a network that joins all the outlets of a certain organization across different cities. Lastly, there is Wide Area Network (WAN) which connects computer and devices which may be very far from each other, as in the case of the internet.

Due to the grouping of more than one computer, any unauthorized access to the system can expose much information. Similarly, an illegal penetration into the system lowers the confidentiality and authenticity of the information and the data shared or stored.

Resultantly, developers have come up with strategies that will guarantee the security of data in a network. This is through security measures such as firewalls, multilevel security system, and biometric authentication, among others.

The main aim of Network Security is to ensure confidentiality, integrity, availability, and reliability of the data stored therein by preventing unauthorized intrusion. In case a network is not secure the following can happen:

- Leakage of sensitive information which can be used to defame or, generally used against the correspondents.

- Any unauthorized person can introduce Trojan and viruses to the system. When a single component of the network is affected, chances are that any information that will be sent from the device may have the viruses which will eventually affect the whole network system.

- In case, unauthorized and malicious person gets control of the network, the saved or shared data may be altered or deleted, which can cause chaos in the corresponding institution or organization.

Short Introduction to Cyber Security

With increase in use of technology there are malicious people who will try to use it against you by gaining access to the data illegally. Most of businesses', governments', and personal activities are being carried over the internet. Businesses are sharing data from one station to another through internet or store the same in the cloud storage. The governments have initiated online processing and application of documents. At a personal level, many people are sharing information regardless of their geographical location.

Therefore if any person gets access to any of the network, there are chances of exposure of personal and sensitive information. Cyber-attacks have been a great threat to many individuals, banks, institutions, organization, and even governments.

There may be loopholes in any network, particularly when using WAN. As a result, cybersecurity is the measures placed to secure internet-connected devices, that is, hardware, software, and data, from attacks.

Attacks, in this case, means unauthorized intrusion and access of data which one is not privileged to see.

The main activities that an intruder can conduct are:

- Accessing information illegally.

- Deleting stored information.

- Modifying the data.

Cyber-Attacks and Cyber Hackers

Cyber-attack is a term that refers to any malicious intrusion that gives unauthorized person access to your information and data. On the other hand, cyber hackers are the persons who perform those criminal acts of gaining access to certain, system, data or information illegally and may decide to modify the data. Similarly, they can block the genuine user from accessing the data until the subject meets some conditions that the hackers will place.

Hackers can do the following to any data they gain access to:

Alter the Confidentiality of Information: Some information or stored data may contain confidential information about particular person. Hence unauthorized access will breach the confidentiality of the information.

Interfere with the Integrity of The Information: The second activity that hackers can do is modifying the information to work in their favour or the person behind the scene.

Interfere with Availability of the Information: Hackers can delete any information at their disposal or block the authorized from accessing the information.

The History of Cyber Security and Cyber War

Each and every day, cyber-attacks are becoming sophisticated due to increased technological advancements. As new technology emerges, hackers will eagerly work to find loopholes. Generally, there are types of cyber-attacks namely:

- Unpatched software.

- Phishing attacks.

- Trojan horses.

- Network-traveling worms.

- Advanced persistent threats.

- Denial-of-services.

Hacking and cyber-attacks have been there for long. In history, there are several occurrences of cybercrimes as recorded below.

The Morris Worm: So far, this has been regarded as the first cyber-attack which occurred in 1988. However, it is believed that Robert Tappan Morris, Cornell University graduate, started the activity with a good intention, but it ended on the wrong side. The primary intention of the program he had developed was to establish the number of computers connected to the internet. However, in the process of making the program more accurate it got installed in a computer forcefully even if the computer has had a similar installation. Eventually, it caused crashing of systems, and it is regarded as the first Denial of Service attack as it affected 10% of entire internet-connected computers of that time. Consequently, Morris was charged with violation of computer fraud and abuse act and was sentenced three years of probation and community services plus fines.

LA KIIS FM Porsche Contest: This was an intentional cyber-attack that occurred in 1995 and conducted by Paulsen. LA KIIS FM was awarding the 102nd caller a Porsche. As a result, Paulsen used his hacking abilities to claim the victory. He illegally intruded the channel's phone network and blocked their ability to receive calls; resultantly he became the 102nd caller thus emerging the winner. However, he was eventually arrested and sentenced to five years in prison.

2002 Internet Attack: In 2002, the first cyber-attack that affected the internet was recorded. The event took about an hour having targeted 13 Domain Name System (DNS) root servers. Though the Denial of Service did not affect many people, perhaps it sustained for long it could have shut down the entire internet. So far there is no recorded similar sophisticated attack.

Church of Scientology Attacks: In 2008 a group referred to as Anonymous targeted a website owned by Church of Scientology as a Denial of Services attack. The group, which was an activist movement against the church, conducted approximately 500 Denial-of-Services attacks to Scientology website in one week. Resultantly, a Jersey teenager was sentenced two years of probation plus fines for the crime.

2013 and 2014 Yahoo Attack: Though the attack was revealed in 2016 while in a negotiation process, Yahoo was a victim of cyber-attack in 2013 and 2014. In the attack, approximately 500 million accounts were compromised in 2014. A similar attack was also recorded in 2013, where one billion user accounts were compromised. The attacks affected the value of Yahoo with approximately $ 350 million.

2014 JP Morgan Chase: In the attack, hackers gained access to names, addresses, phone numbers, and emails of approximately 76 million households and 7 small businesses accounts. However, the hackers were unable to retrieve social security numbers and passwords.

2016 Adult Friend Finder: In 2016, approximately 412.2 million accounts were attacked, exposing names, email addresses, and passwords for accounts spanning over 20 years. The hackers took advantage of weak password protection, SHA-1 hashing algorithm; as result, even passwords were exposed by the time the attack was discovered.

2017 Equifax Cyber-Attack: In 2017, hackers managed to attack approximately 143 million user accounts exposing sensitive and valuable data. The leaked data included birth dates, social security numbers, addresses, driving license numbers and some credit card numbers. Equifax announced publicly in September that they were responsible for the attack.

2018 Exactis Attack: In June 2018 unknown marketing firm leaked approximately 340 million records which made it double effects of Equifax attack. However, it was established that the cause was weak cyber-security, which made the database accessible to any hacker thus revealing personal information stored therein.

Cyber Security Certification and Laws

Regarding cybersecurity, there are different and distinct areas of specialization which include law of crime, liability, compliance, contracts, and policies. All learners are equipped with techniques such as credible preparation, giving reports in forensics, incident response or other investigations.

Global Information Assurance Certification is always awarded to learners who learn and can conduct cybersecurity processes and ensure security of firms from external attacks. Other entities in cybersecurity certification are HIPAA, GLBA, FISMA, GDPR, and PCI-DSS.

Cyber Security Engineering

Cybersecurity engineering refers to securing information, professionally, by performing activities such as designing, establishing and implementing safe network solutions to reduce the vulnerability of a network from an attack, hacking or illegal intrusion.

Most organizations hire cybersecurity engineers to, frequently check on the vulnerability of their networks from attack. The engineers can work in isolation or groups depending on the task. Therefore, cybersecurity engineers should possess expert-level understanding of how different networks operate and how activities in a network happen under normal circumstances. This will help them in identifying abnormality, which in most cases is an identifier of an attacked network.

Additionally, the engineers will have to conduct penetration tests where they can be termed as white hackers. White hat hackers work to establish the chances that a malicious black hat hacker has to intrude into the network. Therefore, an engineer in an organization has the role of testing the network, computers and application for vulnerabilities.

Similarly, all the engineers should be equipped and aware of the established laws in cybersecurity. They should comply with Gramm-Leach-Bliley, HIPAA, GDPR, PCI-DSS, data breach notice laws, and other regulations. Of equal importance, they should be aware of agreements, insurances, outsourced services, private investigation services, and also cloud computing issues.

Cryptography

Cryptography is the solution for cyber-attacks that are achieved through coding the information to make sure only the privileged have the ability to read and process it. This can be achieved through encrypting the data as the name, cryptography, 'crypt' means hidden all vaulted message while *'graphy'* means processing or writing.

The main aim of this branch of computer science is to ensure that the shared message is hard to decipher unless one is privileged to do so. Some encrypted message will require passwords to open them; this technique means even if the information lands in the wrong hands, it will be useless unless the recipient has the passwords. The set passwords are mostly, unique number such as national identification numbers, passport number or number that is solely associated with the recipient.

The primary goals of cryptography are:

- Ensure that the confidentiality and privacy of the data are maintained.

- Ensure that the integrity of data by making sure it remains original.

- Ensure the authentication of the sent or received data. Recipient or sender can confirm the identity of the correspondent.

Cybersecurity in the Era of AI and ML

Human beings have been regarded as more intelligent than machines, but why? The human being is able to perceive, that is taste, hear, sense, smell, and see. That is why whenever one fails to do things logically the politest response to search a person is *'Use your common sense.'* The main reason behind this is that human being has the capacity to deduce meaning from what they can perceive. However, for long computers have been regarded as having no IQ, it could not respond to climate changes or sense the mood of the day it just does as instructed.

However, artificial intelligence is the features that computer industries are incorporating in the computer to make them have equal or greater than that of human 'Intelligence Quotient.' Therefore, artificial intelligence is the characteristic of the computer that will enhance its perception and response and stop being a 'dumb machine.'

On the other hand, machine language refers to the language that a computer understands commonly referred to as codes.

Artificial Intelligence in Cybersecurity

Artificial intelligence has been established to stop the computer from being a dumb machine but an intellect machine. As a result, AI has been implemented by many organizations and also in combination with machine language it offers tools that support cybersecurity

How Artificial Intelligence and Machine Language can Anchor Cyber Security. Following are the ways that Machine language can anchor cybersecurity.

If an organization manages to detect cyber-attacks in advance, they will be able to counter the adversary from achieving the same. Machine Language as part of artificial intelligence has proven to be effective in detecting cyber threats by analyzing data. The analysis helps in spotting out of any threat and vulnerability of your information.

Through usage and adaptation of algorithms based on the data received, machine language enables the understanding of consequent improvements that are required. Through this feature, machine learning helps in proper prediction of threats by making a distinction between normal and abnormal behaviours. Therefore, when the adversary decides to launch the attack it will have been detected earlier or detected immediately, lessening the effects. Similarly, it will show specified threats which human efforts cannot detect making the computer more intelligent.

Since hackers will always keep advancing in their tactics, there are many threats to deal with daily, which at times can exceed human capabilities. However, AI is a conventional technology and will help in anchoring cybersecurity by adapting to the advancements.

On the other hand, all of the efforts should be made to make the network naturally secure through the following activities.

Using passwords and authentication protection: For any hacker to cause damage or threat to your network, they must have the access first. However, passwords have become so fragile in recent days regarding cyber security. The main causes of weakness experienced in regard to passwords are:

- Most people use similar passwords across all the accounts.

- Using similar passwords for a long time.

- Registering some sites which they are not sure about.

- Being so curious until they later learn it was bait.

In order to make sure that you are always safe at a personal and organizational level you should do the following:

- Have different passwords for different accounts.

- Update your password with time to prevent any hacker who may have reached your previous password from accessing the account and perhaps conducting denial-of-service hacking.

- Do not follow links that you are not sure about. Some links will direct you to accounts that have your sensitive data, and before you note it you will already be a victim.

On the other hand, developers are using Artificial intelligence to develop biometric authentication to replace numeric and alphanumeric passwords. Consequently, you will not need to update your passwords and using different passwords. Such features include face recognition and fingerprints.

Nevertheless, some people have referred to some of these features as irritating due fail to work perhaps when the user has changed the hairstyle the face recognition, ay take time to recognize. As a result, developers are working to innovate more intelligent devices that will overcome such challenges. According to the recent research there are one in a million chances of fooling AI particularly, through biometric authentication.

Using techniques of detecting phishing and prevention controls: Phishing is a technique that hackers use by pretending to be a genuine service provider in order to extract your security details. Researches show that one in ninety-nine unrequested email is a phishing email. Therefore, you should open only the emails you were expecting. Otherwise, a hacker can gain authority over your account.

Additionally, AI-ML has developed ways of preventing and deterring phishing attacks.

AI-ML can detect and track more than ten thousand phishing sources and respond quicker than a human being could. Similarly, Artificial Intelligence makes it possible to differentiate between legitimate and fake website in seconds unlike in human being.

Vulnerability Management: Vulnerabilities is the chances that your network has in getting attacked. However, there can be thousands of unique vulnerabilities identified yearly; this means human effort to identify such is close to impossible. However, Artificial Intelligence proactively searches for potential vulnerabilities in the system. Artificial Intelligence looks for potential vulnerabilities by combining numerous of factors such as debates in dark webs and patterns used.

As a result, AI combines the collected information to establish when and how the threat can make its effectiveness to targets.

Network Security and Artificial Intelligence: To achieve a maximum network security level, there should be creation of security policies and understanding the network topography of the given organization. While these activities can take a lot of time, Artificial intelligence learns the network traffic and suggest the most appropriate security policies.

Artificial intelligence in behavioural analytics: In most cases, the behaviour of a given network can be used to detect if there are an intrusion and modification. However, it may take time for a person to learn even the most minor activities that happen in system. As a result, implementation of AI can help in analysis of behaviour through observing pattern and algorithms, and thus it can easily detect any abnormality at its onset.

The Application of Cyber Security in Government, in Life and At Work

Hackers, particularly the black hat, are mercenaries who get access to your information; they use it for their own gain. Therefore, to avoid being victims of higher institutions and individuals with sensitive information are the most vulnerable to the attacks. As a result, cybersecurity knowledge can be used to save them from the fate. The following are the activities that will ensure a promised security of network and data at large.

1. Plan for a Unique Addresses authentication. A unique authentication makes it hard for a hacker to imitate you. For instance, AI helps in creation of biometric authentication where there is only one chance a million of personification.

2. Understand all the authentication limitations by understanding cybersecurity laws and policies.

3. Ensure your authentication can allow but also track the different geographical locations.

4. Focus on standards and results of the security measures for promising security.

5. Choose authentication processes that are not effective but not complex. The complexity of a system will make it hard for you to note abnormalities.

6. Conduct penetration tests to establish any vulnerability that your network may have for attacks.

7. Ensure privacy and installation of firewalls in your network. Similarly, use AI technology to detect phishing emails.

CHAPTER 7: LIST OF 50 KEY TERMS USED IN MACHINE LEARNING INDUSTRY AND WHAT YOU SHOULD REMEMBER

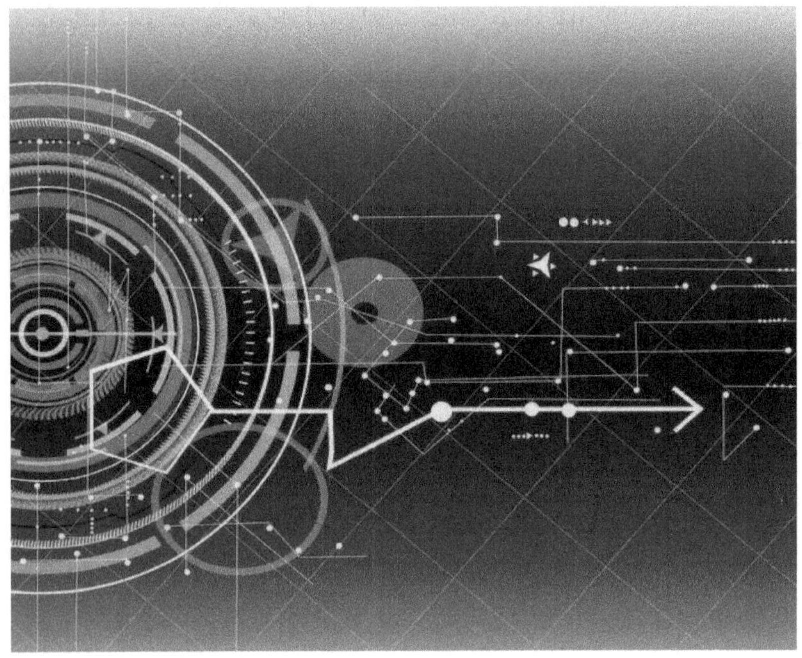

"There's a lot that machine leaning can't do that humans can do very, very well"
Diane Greene

Technology is now ruling almost every aspect of our lives. As discoveries are made nearly every day, human beings may soon have their lives controlled by technology. The concept of machine learning is an exciting field in computer science. Machine learning applies big data to develop sophisticated computer systems that can process the data and use results to predict the future. In machine learning, the computer themselves learn to process data without any prior programming. Machine learning applies specific procedures to establish patterns from a particular big data. Machine learning procedures enable industries to apply big dataset to predict future trends and manufacture products that comply with these predictions. Some of the areas using machine learning include medicine, banking, insurance and medicine. Various products have used machine learning. These include Facial Recognition used by Facebook, Siri and Cortana used for voice recognition, PayPal uses data to detect any fraud etc. Machine learning is under the field of artificial intelligence. Artificial Intelligence is the ability of machines to behave like human beings, especially in the area of knowledge and tackling problems.

Terms used in machine learning

Any field of study has various terminologies that are used. Machine learning utilizes a variety of vocabularies that are used to refer to various concepts. The following are the key concepts that are applied in machine learning:

- **Classification**- This is one of the areas of supervised learning. Supervised learning entails offering the computer the yearned for input and output data. The data submitted is easily categorized. In this regard, there are such categories like binary and decision tree. In binary classification has both a two classification and multi-classification categories. For instance, the computer can classify emails as either junk or non-junk.

- **Clustering**- This is a type of unsupervised learning that categorizes data per specific traits. You can apply the clustering method when classifying clients into various groups based on demographic knowledge. This approach will help you to target them with the right marketing messages. Multiple fields can apply the clustering method. For instance, libraries can use clustering to organize books according to titles or topics. Grouping can assist you in making an informed decision. For example, if you want to start a shoe business in a specific town, this method may assist you in understanding a location where you're likely to find clients for shoes.

- **Regression**- The term regression is used to indicate the relationships that exist between various kinds of information. Regression is under supervised machine learning. Regressions can also be used to check the correlations that exist between different sets of data. For instance, you can use the various features of a property to predict its value in real estate. There are different kinds of regression. These include Simple linear, support vector, random forest and support the tree. Within the machine learning field, regression is essential because you can apply the available data to predict the future. For instance, in medicine, a physician can use regression to predict the likelihood of a patient recovering fully from cancer.

- **Deep learning**- This term is synonyms with machine learning. Deep learning can handle big data which is unstructured the same way human beings process information. The amount of information involved is so vast that it requires many computers to handle. Deep learning, assist in data analyses and prediction of future trends. The primary application of deep learning is to detect fraud in the banking industry. There are various fields which utilize deep learning. For example, in commerce, deep learning can be applied in Chatbots and instant bidding. Deep learning is associated with particular terminologies, including:

> ❖ **Chat box**-This program can participate in a conversation and respond accordingly just like human beings.
>
> ❖ **Real-time bidding**-This algorithm assists you in purchase ads.
>
> ❖ **Recommendation engines**- Recommendation engines use the past behaviour of clients to target them with the right message. These engines use algorithms that filter data and select the one that is relevant to particular users.

- **Logistic regression**- This form of regression establishes a link between input information to forecast the resultant output factors.

- **Activation function**- The activation functions are used on layers and enable neural networks to establish complicated choice borders. **Neural Networks**- These networks resemble deep learning. They help create different layers of neurons that increase the knowledge of information derived from machines to offer error-free evaluation. Neural networks comprise of nodes which are activated to indicate the amount of significance of specific data. The neural network has three essential layers which include input, hidden and output layers.

✓ **Natural language processing**- This subfield of machine learning helps to analyze human speech. Through natural language processing (NLP), the computer can know, examine, control and even originate human language. Natural Language Processing helps machines to retrieve information, translation, simplify information, summarizing of data etc. There are various challenges to natural language processing. These include learning how to process extra-segmental aspects of language, the language used within specific cultures, and ambiguity in sentence construction. In machine learning, computers transform natural language into a form that is recognized and processed by a computer. Through the use of syntax and semantics, processors can analyze and interpret literature. There are various kinds of natural language processes. These include sentiment analysis, machine translation, speech recognition, semantic search etc.

✓ **Text classification**- The aim of this kind of machine learning is to forecast a label of a book. In case the documents are in a list, this method enables the system to label them according to their relevance. For instance, through text classification, you can either have spam or non-spam emails.

✓ **Sentiment analysis**- This procedure of machine learning is intended to evaluate the opinion of a client towards a service.

- ✓ **Document summation**- This form of machine learning helps to produce brief documents from long texts.

- ✓ **Speech recognition**- This is a form of machine learning that transforms audio information into textual one.

- ✓ **Machine translation**-through this machine learning task; a text is instantly translated from one language to another.

- **Machine vision**- This form of machine learning assists machines to recognize photos and interpret them. This type of machine learning is especially helpful in the medical field, where it can be applied to determine whether a patient has cancer. You can also use this technology to evaluate the status of traffic on the road. This technology is essential as it assists in detecting fake IDs and identifying whether the person photographed is a human being or robot. Machine vision is also useful in helping to determine the criminal found on a CCTV camera. Various aspects are essential in machine learning. These include:

- **Image classification**- Through this task, you teach a model on how to recognize the contents of a photo. For instance, you can train a design to identify an area with a disease in an x-ray image.

- **Object detection**- This machine learning model helps to train the machine to discover an object using predefined features.

- **Image segmentation**- You can teach your model to classify images using specific qualities.

- **Saliency detection**- This model will help you determine the areas of an image that viewers will pay their most attention. Saliency detection enables you to come up with high-quality adverts that target specific people.

- **Machine Learning Engineer**- This is a professional who uses data to develop computer designs that are capable of learning just like human beings. These professionals know computer technology, especially computer science and data application.

- **Machine learning**- In computer science, machine learning is a type of artificial intelligence that focuses on creating computer systems that can analyze data and give predictions. Machine learning help organizations to process big data and use the analysis to come up with products and solutions that many people want.

- **The supervised learning**- This term implies that a computer application is tutored to process specific data. The knowledge that a computer program derives from this learning enables it to come up with error-free decisions based on the information provided.

- **Unsupervised learning**- In this form of learning, the computer systems train themselves by recognizing the data given. Through unsupervised learning, models apply procedures that use unlabeled data. This form of learning assists machines to classify information according to specific characteristics. One example of unsupervised learning is clustering. Through clusters, computer models organize data in different groups. Each cluster or group has similar traits.

- **Decision tree**- This type of machine learning enables a computer to make a decision based on a graph that appears like a tree. The decision tree model allows the system to understand the repercussions of a specific decision. A decision tree is one classic example of how you can visually represent a procedure.

- **Model**- This is a representation of reality using a mathematical framework. To produce the machine learning model, you must feed your machine with a learning algorithm.

- **Generative model**- This form of the model enables the machine to produce information when you hide specific variables. Generative model in machine learning can be used to allow the business to evaluate what it requires to be efficient in offering its services.

- **Discriminative model**- This type of model is applied in machine learning to fashion the dependence on a parameter over another one, for instance, variable Y on variable X.

- **Dataset**- When you want to develop a machine learning system, you require data that you can either collect yourself or get it from particular sources. This kind of data that is applied when developing or examining an ML system is referred to as dataset. There are three forms of data set. These include training data, validation data, and test data. The training data is used to teach design. The dataset on the other hand, helps machine learning to discover patterns or decide the significant attributes for prediction. This validation design is applied to engage with different models and determining the best for your case. Specific test data is used to indicate behaviour on unused data.

- **Reinforcement learning**- This kind of learning enables the computer algorithm to play a match to reap high profits. The computer algorithm tries various games and identifies the one that enhances high rewards. The games that are commonly used under reinforcement learning include chess or Rubik's Cube.

- **Overfitting**- Overfitting implies a negative effect that comes as a result of developing a model on insufficient data. The model is biased and may not offer you accurate predictions. For instance, if you visit a butchery several times for goat meat but you don't find it, you may be annoyed. However, some so many clients are satisfied with beef that the butchery offers. In this case, you're a single person who is disappointed with the butchery's services compared to the majority of clients satisfied with its beef. In this regard, the model that you've developed is biased and does not provide with accurate and factual information.

- **Algorithm**- These are procedures that are applied in machine learning to solve problems. Algorithms help a system to process and analyze data to predict the future. There are four main kinds of algorithms. These include supervised learning, unsupervised learning, semi-supervised learning and reinforced learning. These algorithms are explained below:

- **Supervised learning**- This type of algorithm applies labelled data to learn the functions that convert input data in output data. There are two main types of supervised data. These include regression and classification data.

- **Unsupervised learning-** This form of procedure only has only input data. The two main kinds of unsupervised data include clustering and association rule learning.

- **Semi-supervised learning-** This is a combination of labelled and unlabeled data.

- **Ensemble algorithms-** This comprises of many models of weaker models with individual training to have a big prediction when they work together.

- **Training-** In machine learning training data is the initial information that helps the computer to learn to apply sophisticated technologies like neural networks to generate desired outputs. The training data comes at the beginning of the process of machine learning. After that, it is followed by validation and testing data. Training data is used to assist the computer in understanding how to process complex information, just like what happens in our brains.

- **Target-** This is what you intend to predict when you feed the computer with data.

- **Features-** These are conditions that cannot be changed that you put in your computer. Features are the main factors that are used by prediction designs to make forecasts. When evaluating features, you need to determine their attributes and dimensions.

- **Label-** A label is the real result of your target. Labels indicate the class where the information belongs. This means the different clusters of information with certain familiar features are put together for labelling.

- **Overfitting**- In machine learning, generalization is an essential factor that helps the system to predict the outcomes using the training data accurately. To enable the system to generalize explicitly, you need to ensure that the training data used is of high signal to noise ratio. Without this accuracy, generalization will not be useful in providing accurate forecasts. A design is considered overfitting when it accurately fits in training data but is terrible in generalizing new data.

- **Regularization**- This method is used to approximate the best complexity that can be undertaken by machine learning so that to discourage the system from underfitting or overfitting issues. To reduce the independent of the design, various penalties are included in different variables of the model.

- **Machine learning Task**- This is a mixture of information with features and target. There are various kinds of machine learning tasks, including feature selection, clustering, regression, and others.

- **Prediction**- Based on specific features, the machine learning design can forecast a target value.

- **Hyperparameter**- This is a factor whose value is determined before machine learning. Various designs need different kinds of hyperparameters. However, there are those designs that do not need hyperparameters. The salient features of hyperparameters include: manually run, they are applied to assist in approximating design parameters, and they help in enhancing a model's operation.

- **Accuracy**- This is an estimation of absoluteness of a classification model. You can measure it by dividing the value of classified instances by the number of classified instances.

- **Activation function**- In deep learning activation function determines the absoluteness and the effectiveness of training design. Activation functions are represented mathematically and are applied in establishing the results of a neural network.

- **Active learning**- In this kind of machine learning, the model determines that information that it should train from. This kind of learning is essential to label data is dear to access.

- **Area Under the ROC Curve**- The initials ROC stands for Receiver Operating Characteristics. This strategy is applied when analysing the functioning of a classification design. This design utilizes a mixture of false and actual favorable rates.

- **Bagging**-Through bagging or bootstrap aggregating, you can enhance the accuracy and power of machine learning designs as applied in regression. Bagging assists in getting rid of overfitting.

- **Baseline-** This is a procedure that uses simple statistical methods and trial and error approaches to form predictions. A baseline uses basic statistics to develop forecasts of datasets. The baseline information helps users to compare the output from machine learning and the expected results.

- **Base learner-** This is a learning procedure that develops designs that are mixed by an ensemble learning design. An example of a base learner is a decision tree explained above.

- **Bias metric-** This is the value of discrepancy between what is forecast and the accurate value for recognized. There are two forms of bias metric: low bias and high bias. Low bias implies that all predictions are right. High bias, on the other hand, suggests that your design is underfitting, and you're applying an erroneous model to perform your task.

- **Deduction-** This is a method that is used to solve problems by using testing theory through observation. It uses a top-bottom approach.

- **Epoch-** This means the frequency with which the procedure deciphers all the information in a dataset.

- **Induction-** This is a problem-solving procedure that applies a bottom-up approach. You use observations to develop theories.

- **Instance-** This is a subset in a dataset. You can also use the term observation to refer to an instance.

- **Precision-** This is the ability of your design to measure positive observations. It predicts the frequency with which what is forecast is right. In case the expense of false positives is vast, precision comes in hand. For instance, if a system provides negative results when testing whether a patient has cancer, the patient will keep on incurring costs to determine the problem.

- **Artificial intelligence-** This is a subfield of computer science which is concerned with the task of creating computer systems that can collect data and analyzing it for decision making.

- **Data mining-**This is the process of analyzing big data and establishing patterns that can forecast the future.

- **Chatbot-**This is a machine program that is capable of participating in a conversation with human beings.

- **Data-** This is any information which is transformed into a digital type.

- **Genetic Algorithm-** This approach solves problems by aping the biological processes of natural selection and evolution. The design uses a random methodology to choose a pair from the universe that can act as male and female. The chances of selecting active pairs are very high. The genetic algorithm applies three kinds of rules to form a generation from the current population. These include selection laws, crossover rules and mutation laws.

- **Heuristics-** This method is applied in solving problems when the main approaches are slow. It mainly uses experiments, trial and errors, and analysis. The main advantage of heuristics over the other methods is that it is quick.

- **Natural Language Generation-** In this form of machine learning; designs are created that are capable of producing human-like language.

- **Optical Character Recognition-** This is a computer program that captures images of a text and transforms them into computable readable text.

- **Black box model-** The black-box model is a design that you can't understand by reading its parameters. The model's internal attributes are hidden.

The field of machine learning is evolving at high speed. In case you're interested in this area, you're likely to encounter various terminologies that are used. This article has highlighted and explained the critical terms used in this body of study.

CHAPTER 8: DATA MINING

In this chapter, you will learn about the essential aspects of Data mining. We will talk about data mining concepts, techniques, and methods. We will also discuss how data mining is necessary for business analytics, its implementation process, as well as its importance in machine learning. Read on to find out!

What is Data Mining?

Data mining can be described as the process of sorting through vast volumes of data to identify patterns and find relationships to solve issues or problems through analyzing data. Data mining enables business enterprises to predict what is likely going to happen in the future.

It is used by companies to transform raw data into meaningful information. Businesses get to learn more about their customers to develop effective marketing strategies, reduce the production cost, and increase sales. Data mining relies on warehousing, effective data collection, and computer processing.

Data mining processes are used to create machine learning models that can be used to power applications such as website recommendation platforms and search engine technology.

How does Data Mining work? It involves analyzing and exploring large sets of data to establish meaningful patterns and trends. Data mining can be used in credit risk management, database marketing, fraud detection, and spam email filtering. We will discuss more the uses of data mining later on in the article.

Having known what Data Mining is, next, we discuss its concepts, techniques, and methods. Read on!

Data Mining Concepts

Creating a data mining model is part of bigger processes that involves everything from asking questions about the data in question to developing a model that can help analyze and answer the questions. The following six steps can clearly explain the process.

i) Defining the Problem

This is the first step in the data mining process. It includes analyzing business necessities, defining the core of the problem, defining the specific objectives for the data mining process, and defining the metrics by which the model is going to be evaluated.

To answer those questions, you will have to do a data availability study to determine the needs in relation to the available data. If the data does not support the user's needs, you will have to redefine the project. You need to put in consideration methods in which the results of models can be merged in key performance indicators that are used to determine business progress.

ii) Preparing Data

This is the second step in data mining. It is the stage where data is consolidated and cleaned. Data, in most occasions, can be scattered across an organization and stored in various formats. It may contain abnormalities like missing or incorrect entries. For instance, data might indicate that a customer purchased a product before the product was even introduced in the market.

Cleaning data isn't just about doing away with bad data or filling up missing values, it is about finding the hidden correlations in data, identifying most accurate sources of data and establishing which columns that are most fit for use in the analysis.

For data mining, you are typically working with a large set of data and can't determine each and every transaction for data quality. This will, therefore, need you to use data profiling and automated data filtering tools. Some of those tools are Microsoft SQL Server and SQL Server Data.

However, it is essential to note that the data used for data mining doesn't need to be stored in Online Analytical Processing cube.

iii) Exploring Data

This is the third stage in the data mining process. In order to make appropriate decisions when you create a mining model, you must first understand the data. Some of the exploration techniques are calculating the maximum and minimum values, calculating standard deviations and mean and exploring data distribution. Standard deviations and other distribution values can give useful information about the accuracy and stability of the results.

You can employ tools like Master Data Services to establish available sources of data and determine data mining availability. You can also use tools Data Profiler and SQL Server Data Quality Services to analyze your data distribution and repair errors like missing or wrong data.

After defining your sources, you can combine them in a Data Source by applying the Data Source View Designer in SQL.

iv) Building Models

The fourth step involves building a mining model. You will apply the knowledge you amassed in Exploring Database stage to help create and define models.

To define the columns of data you prefer to use, create a mining structure. The mining structure is connected to the source of data but doesn't have any data unless you process it. The information can be used in any kind of mining model that is based on the structure. Before structure and model processing, a data mining model is just a container that outlines the columns used for input and the parameters that tell the algorithm how to process the data.

You can also use parameters to adjust algorithms. Additionally, you can apply filters to data so as to use a subset of it, creating different results. After taking data through the model, the mining model object contains summaries that can be used for prediction.

It is important to note that whenever data changes, you should update both the mining model and the mining structure.

v) Exploring and Validating Models

This stage involves exploring the mining models, build and test their effectiveness.

Before deploying a model in a production environment, it is prudent to test how well the model performs. When building a model, you sometimes can create multiple models with various configurations and test all of them to determine which one yields the best solution to your problem.

To explore trends and patterns that the algorithms discover, use the viewers in Data Mining Designer tools. You can test how good the models create predictions by using applying tools in the designer like the lift chart and classification matrix. To determine whether the model is significant to your data or not, use the statistical technique to automatically create subsets of the data and the model against each subset.

vi) Deploying and Updating Models

This is the final step in the data mining process. This stage involves the deployment of models that performed best in a production environment. With the mining models in place, you can perform various tasks depending on your needs. Below are some of those tasks;

- Use models to create predictions that can later be used in making business decisions.

- Creating content queries to retrieve rules, statistics, or formulas from the model.

-Use integration services to create a package that the mining model is used to sort incoming data into multiple different tables intelligently.

Data Mining Techniques

Today, there are several data mining techniques used in data mining projects. We will examine those techniques in the following subsection.

Association

In this technique, a pattern is discovered based on the relationship between things in the same transaction. This is why the association technique is often referred to as a relation technique. This technique is mostly used in market analysis in order to identify the most frequently purchased set of products.

Retailers are using association techniques to discover customers buying habits based on historical sale data. This might help them establish that customers always buy a pack of crisps when they beers and they can, therefore, put crisps and beers next to each other in order to save the customer's time and in the long run, increase sales.

Classification

This is a data mining technique based on machine learning. It is used to classify each item in a set of data into one large predefined set of groups. The classification technique puts into using mathematical methods like linear programming, decision trees, neural network, and statistics. In classification, you develop software that can learn how to classify data items into groups.

Clustering

It is a data mining technique that makes use of a cluster of objects that have similar characteristics using an automatic technique. It defines the classes and categorizes objects into each class. For example, in book management in a library, there is a wide range of books each talking about different topics. The challenge in this situation can be keeping those books in such a way that readers can pick several books in a particular topic without a struggle. By using the clustering technique, you can keep books that have similarities in one shelf or cluster and label it with a relatable name. So, if readers want to pick books in that topic, they will only have to go to that shelf rather doing rounds in the entire library.

Prediction

Just as the name implies, the prediction is a data mining technique that discovers the link between independent variables and the relationship between independent and dependent variables. For example, the prediction technique can be applied in the sales to predict profits for the future if the sale can be considered as an independent variable with profit being the dependent variable. Based on the previous sale and profit data, you can draw a fitted representation curve that is used for profit prediction.

Sequential Patterns

This is a data mining technique that looks to identify similar patterns, trends, or regular events in transactional data over a business period.

In sales, establishing patterns can help businesses identify a set of items that customers buy together at different times. The businesses can then use the information to advise customers to buy it with better deals basing on their buying frequency.

Decision Trees

This is one of the most used data mining technique because its model is easily understandable for users. In this technique, the root of the decision is a simple question that has several answers. Each answer results to a set of conditions that help you determine the data that you can base your final decision on.

Data Mining Methods

There are several methods of data mining and data collection. Below, we discuss the most common methods of data mining and how they work.

I)Anomaly detection

This can be used to determine something is different from a regular pattern. For example, monitoring gas turbines and how you can detect anomaly to make sure the turbines are properly functioning. Sensors monitoring pressure and temperature are set up to see if anything abnormal is observed over time.

ii)Association learning

This is used to determine which things tend to occur simultaneously, either in pairs or groups. For example, it can be noted that people who buy milk often buy it break, and those that buy diapers, buy baby formula as well.

iii) Cluster detection

This is recognizing distinct groups within data, and the process is called cluster detection. Machine learning algorithms detect significantly worrying subgroups within a set of data.

iv) Classification

Unlike the case in cluster detection, classification deals with things that have labels. For example, spam filters are used to identify differences between content found in legit and spam messages. This can be done by identifying large sets of spam or email.

iv) Regression

This method is used to predict the future basing on the relationships within a data set. For example, the future engagement on the Facebook platform can be predicted based on everything in the user's history like photo tags, likes, infractions with other users, friend requests, among others.

Similarly, another example would be the relationship between income and education level to predict the future of a neighborhood. The regression method allows all the aspects and relationships within a set of data to be analyzed and then used to tell the future behavior.

Data Mining Implementation Process

As you have learned earlier, data mining process the sorting through of large sets of data, relationships and insights that lead enterprises in measuring and managing where they performance is now and predicting where it will be in future.

Large sets of data are brought in from various data sources and may be kept in different data warehouses. Data mining techniques such as artificial intelligence, machine learning and predictive modelling can be involved.

The process of mining data requires commitment. Experts agree that the data mining implementation process is the same, and should follow a prescribed path.

Discussed below are the six essential steps in the data mining implementation process.

1.Business Understanding

In this phase you must fulfill a number of things. First, it is needed you understand your business objectives clearly and discovering what your business requires in order to prosper.Next, you have to asses the situation at hand by finding the assumptions, resources, constraints and other important factors that should be considered. Then, from your business objectives, build data mining goals to realize the business objectives within the current situation. Lastly, a credible data mining plan has to be implemented to achieve both data mining and business goals. You should ensure your plan is as detailed as possible.

2.Data Understanding

This is the second stage in the data mining implementation process.It starts with initial data collection which is sourced from available data sources to help get relevant with the data. Some essential activities must be done including data integration and data load in order to make the data collection a success. Then, the 'surface' properties of the sourced data needs to be examined very carefully and reported. Later, the data needs to be analysed by answering the data mining questions which can addressed through reporting ,quarrying and visualization. Finally, data quality must be analysed by answering essential questions like "Is the data acquired complete?", "Are there any missing values in the acquired data?"

3. Data Preparation.

This process typically consumes about 90% of the period of the project. The outcome of data preparation stage is the data set. Once the available sources of data are identified, they need to be chosen, cleaned, constructed and formatted into the ideal form. The task of data exploration at a greater depth maybe be done during this phase to identify the patterns based on business understanding.

4.Modelling.

To begin with, modelling techniques have to chosen to be used for the prepared data set. Next, the test scene must be generated to determine the quality of the model. One or more models are then created on the data set prepared. Lastly, the models need to be assessed carefully involving all the stakeholders to ensure that created model meet the set business initiatives.

5. Evaluation.

In this second last phase, the model results are evaluated in relation to business objectives drafted in the first stage, new business needs may be raised due to the new patterns discovered in the model results or from different factors. Understanding the business is a vital process in data mining. The do or don't do decision must be made in this phase in order to move on the initiation phase.

6.Implementation

The information or knowledge gained through the entire data mining process needs to be showcased in a way that stakeholders can put it into use when they need it. Basing on the business requirements, the implementation phase could be as simple as creating a report or as complicated as repeating the entire data mining process across the company. In the implementation phase, the plans for deployment, monitoring and maintenance have to be created for implementation and also for future support.

Data Mining application in Business Analytics

Retail, finance, and marketing firms predominantly apply data mining in their operations. It helps create predictable vital information like target audiences, buying frequencies, and customer personality profiles. Therefore, data mining plays a crucial role in business analytics, as you will see in the next discussion.

As A Decision-Making Tool

Competitor analysis, market research, and industry studies are key in the intelligent decision-making process of any company.

In the banking sector, customer data is analyzed to determine customer behavior. While the sector has statistical tools for more automated trend analysis, data mining applications have ensured structures move to more objective analysis.

A relationship management strategy is key for every company. For corporate organizations, CRM has to work towards improving company productivity and better client relations.

Statistical and quantitative marketing techniques have been applied mainly to business analytics. Database marketing has been pivotal for most companies and has been increasing since the evolution of the Internet. Businesses can now predict product marketing models, response, and potential target audience for any product before it even launches.

Data mining has also been applied in market basket analysis. This is a trend in the retail industry that has caught productivity with globalization. This helps design a specific strategy for the layout of the store and marketing strategy of various products and events.

Data mining provides financial institutions with vital information about loans and credit reporting by creating a model for historical customers. Additionally, it enables banks to detect fraudulent card transactions hence protecting the card's owner.

In marketing campaigns, data mining plays a key role. This is because it helps identify customer response. This enables businesses to know which products are on high demand and on those that aren't.

Data mining provides customer response after a marketing campaign. This provides informational knowledge while determining customer groups.

CHAPTER 9: MORE DETAILS ABOUT DATA MINING

"Whenever the price of cryptocurrency is rallying, people start spending a lot more"
Erik Voorhees

In this digital era '' Data Mining" has become a term that is widely used in different circumstances. It is a term that you will frequently encounter in business. This term is also popularly used in cryptocurrency mining and blockchain. But what is data mining? What does it entail? And how does it work? In this piece, we will tell you everything you need about data mining.

Data Mining for Masses

Data mining is best described as a discipline that involves the collection and categorizing data. It is a process in which all efforts are geared towards going through raw data, with the intent of turning into bits of useful information. How is this important? Well, in every business, various transactions are captured by the systems involved in the execution of relevant processes. This leaves a large trail of data in its raw form.

When data is in its raw form, it becomes difficult to utilize it for business improvement. This is where data mining comes in. For example, when you sign up for various programs or even a credit card, you leave a trail of data. Usually, this is meaningful information concerning your preferences, behaviors, spending habits, and interests.

People who sign up for a similar program leave this kind of information in the system. Collectively, it becomes an enormous data tank. Companies have to sift through this data and take the bits of information which could help grow their businesses. For example, if a company wants to increase its sales, there is some information in the large data pool that could be helpful. A data mining process enables the company to obtain useful data for that particular purpose.

Through data mining, companies decipher specific customer's traits and patterns. These patterns are then used to shape up effective business strategies. This is how companies improve their service delivery processes to meet their customers' needs.

Techniques Used in Data Mining

Four common techniques are used in data mining. These are;

1. Regression- best described as predictive.

2. Association Rule Discovery- descriptive.

3. Classification-predictive

4.   Clustering-predictive.

This list is not exhaustive as there are more data mining techniques used across all industries.

Data mining tools

When data is collected, it is held in the company's database. Later, it is retrieved during the mining process. Powerful computing devices are used to access these databases during the mining process.

Besides computers, other tools are required to make the data mining process successful. Since the data is amassed in computer hard drives, software's are some of the tools used in data mining. Why are these tools needed? Data warehouses are vast and hold tons of complex information. For this reason, tools are required to sift through the data and categorize it.

This is a difficult time-consuming task. A group of individuals can't possibly single-handedly extract and analyze all the data manually. Therefore, tools are used to extract as well as convert the data into comprehensible structures. This makes it ready for future use.

Here is a list of top data mining tools and applications:

1. Rapid Miner

Rapid Miner is one of the most coveted tools in data mining. It is made by a company that goes by the same name. The developers of this software used JAVA programming language to write it. It is a powerful tool that provides clients with numerous possibilities in data mining.

Rapid Miner is best described as predictive analysis software. It provides a combined environment for text mining, deep learning, and machine learning. This tool can be used for data mining in different fields. These include business applications, research, medicine, education, training, commercial applications, and application development.

Rapid Miner has three modules. These are;

1.   Rapid Miner studio

2.   Rapid Miner Server

3.   Rapid Miner Radoop

Rapid Miner Studio handles tasks related to prototyping, validation, and workflow. Once the predictive data models are created, the Rapid Miner Server is tasked with operating them. These models are created in the studio module. Rapid Miner Radoop then breaks the models down resulting in predictive analysis.

2. Weka

This is free software, that's popularly referred to as the Waikato Environment. Weka is a machine learning software that was developed at New Zealand's prestigious Waikato University. It performs best in predictive modeling and data analysis. Some of its outstanding features include visualizations tools and algorithms. These are extraordinary features tasked with supporting machine learning.

Weka software is also written in JAVA programming language. It comes with the Graphical User Interface that enhances flawless access to its other features.

With Weka software, you can perform multiple data mining activities. The major ones include data mining, regression, visualizations, and processing. Its working principle assumes that data is organized as a flat-file. Additionally, this software can process data feedback that is returned by the relevant query.

3. Sisense

Unlike the other software, Sisense is in the category of licensed BI data mining software. It's an incredible reporting tool within an organization. This software is a product of a company that goes by the same name. It handles and processes data exceptionally well. Large and small companies can use this tool for all their data mining needs.

With Sisense, a company can combine data from different sources, and still generate satisfactory results. This software is revered for its superb visuals in reporting. Sisense is user-friendly. A non-technical staff member can use it without encountering any technical difficulties. As part of its design, it offers widgets as well as drag- and -drop options.

Widgets come in handy, especially when you want your results to be expressed as pie charts, bar graphs, line graphs and so on.

4. Oracle Data Mining

Oracle Data Mining is classified as proprietary license program. It offers outstanding mining algorithms. These are algorithms that makes it an excellent tool for specialized analytics, regression, and prediction. It is a multifaceted software used by companies to predict, target customers, and even detects fraud.

The graphical user interface of this software offers the drag and drop option. It generates the best insight in various mining activities.

5. Data Melt

Data Melt software is popularly referred to as Smelt. It is classified as an open-source data mining software. The main strengths of this software lie in visualizations and computations. It is more ideal for students, engineers, and scientists.

Dmelt is designed to be multi-platform software written in JAVA programming language. You can run it on all operating systems as long as they are in sync with Java Virtual Machine.

It is suitable for stat analysis, massive data volume analysis, and data mining. Dmelt is mostly used in financial markets analysis, engineering, and natural sciences.

6. Orange

The Orange software is preferred by many professionals in data mining. It's a good choice for data visualization. It is written in Python computing language. This component-based software is endowed with components that are known as widgets. The widgets cover data visualization and predictive modeling.

Unlike many other mining tools, Orange has an interactive user interface that takes out the dullness out of the rigorous data mining process. Also, this software formats data into desirable patterns quickly. One can move data through a simple procedure that involves flipping its widgets.

7. KNIME

KNIME is a product of KNIME.com. Its best functions include reporting and integrating data. KNIME's principle of operation is known as modular data pipeline. This software combines various data mining components with machine learning.

The pharmaceutical industry is popular with this software. It also suits companies that are interested in business intelligence, financial data analysis, and customer analysis. Since it is not a complex data mining use, people without a high technical ability can use it.

8. IBM SPSS Modeler

IBM SPSS Modeler availability is on a proprietary license. This is a product of IBM, which is a reputable global firm. It is a data mining tool that's known for text analytics and building predictive models. One of its exceptional features is the visual interface. Users need not use programming as they are already provided with capable algorithms. This qualifies as a user-friendly data mining software.

It comes with the premium and professional models. The premium model offers insightful additional features for advanced activities.

Data Mining Examples

1. Marketing

Companies are always looking for ways to segments their market, target customers, and increases sales. Data mining is deployed in search scenarios to help a company obtain certain information. This could be the age, gender, and customer habits.

The business then uses this information to their advantage. Data mining categorizes their customers is such a way that they can know who target. Once that's determined, a company comes up with promotion ideas that can appeal to the potential customers in a particular market.

This helps them deploy marketing campaigns to the right people with the hope of getting a massive response. If done right, a company can widen its customer base.

2. Banking

This is another great example of data mining. Banks are extremely interested in understanding the risks in various markets. Such information is instrumental to credit rating as well anti-fraud system. The kind of analysis that banks are out to get may revolve around financial data, card transactions, and purchasing habits.

3. Data Mining Is Essential in Day to Day Operations of Banks. It helps them determine how to get the best response from their marketing campaigns. Data mining also helps identify the offers that customers love most.

4. Service Providers

Finding loyal customers can be difficult for service providers. It is imperative to predict how likely customers are likely to change vendors. This is a type of analysis are accomplished through data mining. For this reason, service providers have a knack for keeping large customer databases.

They collect customer's information, such as billing details, website visits, and customer service records. What follows is giving each customer score in regards to these traits. The score indicates the probability of customer switching to other brands.

Gathering this information is instrumental in retaining these customers who are likely to live. Once the customers with a high risk of living are identified, the company has to come with a way of incentivizing them. The main objective of these efforts is to make the customers stay.

5. Retail Stores and Supermarkets

Business owners who run retail stores and supermarkets are keen to know the taste and preferences of customers. With such information in their hands, they know which items to stock, and in which sizes, and at what price. What's surprising is that this data helps them understand you as a customer, more than you know yourself.

With this data, customers are classified in RFM (recency, frequency, monetary). What does this mean? It means that retail enterprises can identify high spending customers, those who may spend a lot of money one in a while, and those who spend small amounts frequently. Such information is then used to create customer offers.

6. Crime Prevention Agencies

Data mining goes beyond business. Crime prevention agencies use to minimize criminal activities. They seek to determine the trends of criminals. What they will be trying to find out is which areas are most affected, what time does the criminal take activities in most cases and so on.

After thorough analysis, they can accurately predict where and when the crime is most likely to take place. Instead of deploying security personnel haphazardly, crime prevention agencies can pinpoint the areas where more security is required.

Data mining is used in borders. It helps immigration officers to know who to search on border crossings. The parameters will be something the age or type of car being used, how many occupants are it as well as their ages.

7. Medicine

The other brilliant illustration of data mining in medicine. It revolves around patient's records, treatment patterns, frequency of ill health, and physical examinations. Medical practitioners rely on this data to administer better treatment.

Health resources also need to be managed. This entails in examining certain health risks and identifying and predicting illnesses that affect people from a certain group.

8. Tv and Radio

The media is dependent on data mining. The network wants to understand its audience much better. Moreover, data mining is important in scheduling airing of programs. Different audiences have to be targeted in distinct ways.

For instance, radios know what to air early in the morning as folks drive to work. They can tell who is likely to be listening in at that time. The same case applies to viewers. This data is valuable as it helps them decide when to air advertisements on behalf of their clients.

9. Machine learning and Artificial Intelligence

Artificial intelligence and Machine learning are becoming more important in today's world. Both of them are data-oriented. Therefore, data mining plays a huge role in each one of them. The recommendation program is an excellent example of both AI and Machine learning.

 When you buy something from particular websites, other complimentary products are automatically recommended for you by the system in place. How does the system decide on the products to recommend? Data mining is the masterminding process behind all of it. This happens through analysis. It could b through studying your habits of spending, interests, and other traits.

A probability of what you are likely to buy next is established through data mining. Appropriate products are then automatically recommended for you. The same formulae are applied on YouTube, Netflix, and Spotify.

Cryptocurrency Mining and Blockchain

Cryptocurrency has continued to grow in popularity over the last few years. Usually, when users transact, these transactions between various users have to be verified. After the verification process, the transactions are added to what is known as a blockchain digital ledger. This an expanding list of records available to the public.

How Does Cryptocurrency Mining Work?

It starts with a miner. This is a node within the network responsible for collecting transactions and arranging them into blocks. Once transactions are made between users, network nodes collect and verify them. Miner nodes consolidate these transactions into what is known as a candidate block.

This process facilitates the creation of coins from these transactions. This particular coin creation transaction is known as the coin base.

What's There to Gain for Miners?

In the cryptocurrency mining process, miners are issued with a reward each time they add a fresh transaction to a blockchain.

What Tools Are Used in Cryptocurrency Mining?

Cryptocurrency currency mining is similar to other data mining processes. This means that the software's are the main tools used in the mining process. Here are some of the most popular cryptocurrency mining tools:

·   GGMiner

·   Minergate

·   Nicehash

·   Electroneum

·   CGminer

Data mining plays a huge role in today's world. Government agencies and business organizations all over the world rely on data mining, to analyze, categorize, and predict. Data mining continues to evolve, helping businesses better their services, reach more customers, and strengthen improve marketing strategies. It is a phenomenon that the government and businesses can't afford to avoid.

CONCLUSION

My sincere appreciation for making it through to the end of MACHINE LEARNING FOR BEGINNERS: *Easy Guide of ML, Deep Learning, Data Analytics and Cyber Security in practice. Modern approach of Neural Networks, Predictive Modeling and Data Mining with 50 Key Terms.*

I hope that it was educational and stimulating, fluid to read and was able to provide you with the basic knowledge you need to understand the concepts of Machine Learning. By ending this guide, you will be able to understand the role of Machine Learning in influencing human behaviors in different facets of life and what deep impact it will produce in all business sectors.

We have gone through the fundamentals, basics, applications and types of machine learning. The intent of this guide was to provide you practical and effective definition of conceptions that are crucial in understanding Machine Learning.

To enrich this guide we insert a list of 50 most important terms you should know, we hope it can help you to fix the technical words.

You are now familiar of the key terms and applications of Machine learning. We are sure it will help you to find out how it works. With this knowledge you can further learn the more opportunities that exist in such technology.

Ultimately, if you found this book useful and stimulant in any way, a review on Amazon is always appreciated!

Artificial Intelligence and Machine Learning

AI Superpowers and Human + Machine a Visionary Revolution in Finance, Medicine and Business. Find out Top Influent People of the Era with a Modern Approach

Introduction

Congratulations on purchasing *ARTIFICIAL INTELLIGENCE AND MACHINE LEARNING: AI Superpowers and Human + Machine A Visionary Revolution in Finance, Medicine and Business. Find Out 10 Most Influent People of the Era With A Modern Approach*, and thank you for doing so. With the rising world of big data, we are forced to develop effective strategies to help us make informed decisions. All professional fields have become very complex due to the growing demand to find the most efficient ways to extract value from the existing data. By downloading this book, you have taken the first step towards learning how to use big data and how artificial intelligence has made it possible to cope with the naturally outgrowing big data movement. The information that you find in the following chapters is very important as it will help you to learn how artificial intelligence is influencing human behavior and the possible impact in the future.

To that end, this book provides an in-depth overview of artificial intelligence and machine learning, highlighting their historical development and application in various fields including finance, business, and medicine. It covers how artificial intelligence interacts with human intelligence, including the possible partnerships between humans and machines and how each influences the other. People can utilize so much from artificial intelligence and machine learning to improve their lives and enhance their productivity. An interesting concept covered in this book is the AI superpowers across the world, and how they impact the development of AI today and possibly in the future. These include Amit Singhal of Uber, Andrew Ng of Baidu, Elon Musk of SpaceX, and Tesla, among others.

Although there as several options of Artificial Intelligence and Machine Learning books in the market, I am grateful that you considered this one! Please enjoy reading!

Chapter 1: Understanding Artificial Intelligence

"By far, the greatest danger of Artificial Intelligence is that people conclude too early that they understand it"
– Eliezer Yudkowsky

It might not be possible to identify what part of the modern society that has not been affected or influenced by artificial intelligence (AI). Intelligence machines are influencing almost every facet of our lives including business operations and our daily activities with the aim of improving efficiencies. AI is so intertwined into everything we do that it is hard to imagine living without it.

AI is the major influencer of the 4th Industrial Revolution disruptive changes; the changes that will probably challenge human ideas. It is transforming the way we interact and live in an accelerating manner, making our cities to become smarter and help us manage our lives.

Despite the popularity of AI and machine learning, the purpose, and details of the two concepts are not well understood. In this chapter, we are going to discuss the key concepts of artificial intelligence and machine learning.

Defining Artificial Intelligence and Machine Learning

As commonly defined, artificial intelligence is the ability of machines to make decisions and learn in a similar manner as humans. It is commonly known as a branch of computer science that deals with the simulation of behaviors of computer intelligence. Commonly abbreviated as AI, artificial intelligence refers to a computer system that can complete tasks that normally require the intelligence of humans, including recognition of speech, visual perception, decision-making, and language translations. This technology has influenced several consumer products and has promoted breakthroughs in physics and healthcare.

Machine learning is a sub-branch of artificial intelligence, i.e., all machine learning as components of AI, but not all AI count as the latter. It refers to the component of computer science whose aim is to build and leverage existing algorithms in order to establish generalized models that offer patterns and accurate predictions. Such algorithms are usually based on mathematical and statistical optimization. Suppose that your goal is to identify the patterns of a phenomenon using historical data; you may use machine learning algorithms to group the data and analyze the cluster results. With a bit of analysis, machine learning helps in the generation of automatic clusters or categories over time.

History of AI and Machine Learning

Amidst the fact that artificial intelligence and machine learning have been around for centuries, their revelation was not possible until the late 1950s. A series of philosophers, scientists, and mathematicians attempted to explore the concept of AI, but it was not until the World War II when Alan Turing, a British Polymath, suggested to people how to use available information to make decisions and solve problems. Alan was able to figure out and understand the "Enigma" code that was used by the armed forces in Germany to securely send messages. Together with his team, Alan invented the Bombe machine, which helped in deciphering the messages. Turing concluded that a machine that was able to converse with humans could be referred to as an "intelligent" object.

In the early 1950s, John McCarthy, the top computer scientist in the US, organized a conference where the term, "Artificial Intelligence" was invented. Researchers across America became interested in understanding the concept of AI, thus, exploring the subject. Both Herbert Simon and Allen Newell were very instrumental in promoting the field of AI that would later transform the world.

In 1951, Ferranti Mark I, a unique machine, was able to use an algorithm successfully to analyze and master checkers. In turn, Simon and Newell invented an algorithm, commonly known as the General Problem Solver, to solve mathematical problems. Also, in the 1950s, John McCarthy developed the LISP programming language, which was considered a significant part of AI and machine learning. And in the 1960s, the development of algorithms was emphasized by several researchers to address mathematical problems. In the same decade, computer scientists discovered Machine Vision Learning which could be used in controlling robots, and in 1972 the first intelligent robot, WABOT-1, was established in Japan.

However, amidst the global effort to enhance AI applications, computer scientists realized that it was challenging to create intelligence in machines. In order for AI and machine learning to become successful, scientists needed an enormous amount of data. However, the computers that were available at the time were not well-developed to handle such magnitude of data. Corporations and governments lost faith in AI, and in the 1970s and 1990s, scientists experienced a shortage in the funding of AI research.

In the 90s, several giant companies once again developed an interest in developing AI. The government of Japan unveiled the plans to come up with the 5th generation computers in order to advance machine learning. The enthusiasts of AI believed that computers would soon carry conversations, interpret photos, translate languages and reason like humans. In 1997, the Deep Blue, IBM's computer, became the first to beat all computers in holding and interpreting information.

Over time, the exponential gains in the processing powers and storage ability of computers enabled companies to store vast quantities of data. And in the past 15 years, superpowers such as Google, Amazon, Baidu, and other companies have exploited AI and machine learning to a great commercial advantage.

The Goals of AI

The key goals of AI include:

Knowledge Representation and Reasoning

This is a field within AI that focuses on the implementation and designing of computer representations, which are able to process information. AI aims to automate these types of reasoning through the codification of relationships or rules in a way that can be interpreted easily by the computer system. Some of the common uses of knowledge representation and reasoning include a natural-language user interface that aids communication between computers and humans and computer-aided diagnosis that assists physicians to interpret medical records in the form of images.

Automated Planning and Scheduling

This is another subcomponent of AI that deals with the production of automated action sequences that correspond to the measurements, which can be executed using AI systems. The goal of AI planning is to mechanize and automate the generation of any plan based on a set of actions as well as predetermined objectives and goals. One of the examples of AI planning is the self-correcting computer software, automated information gathering, robots that act as autonomous agents, and computerized suggestions.

Natural Language Processing

Another significant goal of AI is the processing of natural language. This involves analysis and generation of language or data that humans can use for computer interfacing. The national language process aims to implement specific computer systems, which can process language data in large quantities. Some of the latest technologies that currently use natural language processing include the Siri app from Apple Company and Google Now.

Computer Vision

The main aim of this AI component is to assess the automation and computerization of tasks that can be performed by human vision and system development in order to process, utilize, or interpret visual data. Computer vision utilizes several applications, including facial recognition, object recognition, automated image manipulation, and video tracking. Computer vision is commonly applied in facial recognition. Other applications are automated image manipulation, integration with virtual reality, object recognition, and video tracking.

Robotics

Another significant goal of AI is robotics, which is about the construction, design, and operation of robots and machines that can replace human tasks. Robotics is a branch of engineering and science that includes electronic engineering, information engineering, computer science, and electronic engineering. Research on robotics is currently conducted to promote military, domestic, and commercial applications. Amazon, one of the largest global retailers, has autonomous robots in their warehouse with the role of keeping the warehouse organized and the operations more efficient.

Artificial General Intelligence

The goal of AI is also to achieve artificial general intelligence, which demonstrates the system's ability to perform intellectual tasks, similar to human beings. AI achieves its AGI goal through integrated systems, including emotional intelligence, cognitive intelligence, and social intelligence.

The Goals of Machine Learning

Machine learning is primarily used for the following output types:

• Unsupervised Clustering

Clustering refers to the unsupervised technique used to discover the structure and composition of a particular data set. It involving clumping data into clusters in order to observe the resulting grouping. Each of the clusters can, thus, be categorized by a set of data points. The common algorithms used to address this goal include hierarchical clustering and K-means clustering.

• Supervised Classification (Two-Class and Multi-Class)

Classification, on the other hand, involves placing data points into a pre-defined category or class. In some cases, classification problems assign a class to an observation or estimate the probability that specific observation falls under a particular class. An example of a two-class classification is giving a class of Ham or Ham to a particular email, in which 'ham' simply means 'not spam.' On the other hand, the multi-class classification simply means multiple classes. In the spam example, it could mean a third class, which is 'unknown.'

- **Regression Analysis: Univariate and Multivariate**

as opposed to a discrete class, regression assigns a continuous response to an observed data. An example is predicting the price of Dow Jones Industrial Average on a particular day. The value could be a number, which would be perfect for regression. The algorithms used to achieve this goal are simple or multiple linear regression, forest or decision tree regression, artificial neural network, Poisson regression, ordinal regression, and nearest neighbor methods.

• Detection of an Anomaly

Although we may assume that data is always sensible and well-behaved, this is never the case. Sometimes we get erroneous data points because of measurement errors or malfunctions. With machine learning, one can easily detect anomalies, which help in providing a measure of quality control. Common algorithms used to achieve these goals are principal component analysis (PCA or support vector machine.

• Recommendation Systems/Engine

Recommendation engine refers to an information filtering system that is meant to make recommendations in different applications, including music, movies, restaurants, books, products, and articles. The common approaches used in the recommendation are collaborative and content-based filtering.

In order to effectively use big data, machine learning tools that are used to leverage different algorithms include:

- Data quality and management
- GUIs used to build models
- Model finding visualization and exploration of interactive data
- Comparisons of models to identify the best one
- Evaluation of automated ensemble model in order to identify the best values

The Relationship Between AI and Machine Learning

Many times, researchers and students get confused in distinguishing between AI and machine learning and identifying the possible relationship. However, machine learning is particularly the study of computer algorithms, which improve with experience. The machine learning algorithms belong to a branch in computer science, commonly referred to as computational learning theory. AI, on the other hand, is a computer system divided into 3 categories: computational philosophy, computation psychology, and computer science.

Artificial intelligence (AI) began as a field in computer science, focusing its objective on solving tasks that people can do. It is approached using different ways, for instance, writing a program that implements standards devised for scientists. Although humans can do a similar task, hand-crafting standards can be a time consuming and laborious activity. AI has several tools that can be used to solve the diverse problems in computer science, including logic, control theory, search and optimization, neural networks, probabilistic methods, and classifiers and statistical learning strategies.

Machine learning is related to AI in that it is a subfield of AI. It is concerned with the establishment of algorithms, which are used by computers to automatically learn models using data. Applications of machine learning are able to read texts and able to analyze the content. Machine learning can also listen to music, make a decision on whether someone is sad or happy, or find pieces of music that suit the mood of a person. Being part of AI, machine learning focuses on continuous data improvement, data mining, and task automation.

Chapter 2: Artificial Intelligence Vs. Human Intelligence

"It has become appallingly obvious that our technology has exceeded our humanity"
— Albert Einstein

Artificial Intelligence and Automation are terms that are frequently used interchangeably. They are related to programming, or robots, and electronic machines that enable us to work more productively and successfully- regardless of whether it is an industrial vehicle assembly or sending a follow-up email.

In any case, it is more nuanced than that. Automation is fundamentally making hardware or software that can do things automatically- without human mediation. The term artificial intelligence was coined by John McCarthy as an engineering science of making smart machines. Computer-based intelligence is tied in with attempting to make machines or software mimic, and in the long run, supersede human behavior and intelligence.

The ideal situation which humankind is moving in the direction of presently is this: Automated machines gather data— AI systems "understand" it. We are looking at two altogether different frameworks that impeccably complement one another.

Automation

Automated systems are the explanation banks record your installments in a matter of seconds, the reason why organizations send mass emails to their clients. It is what enables you to get your consignment shipped and delivered on the same day within hours of ordering.

Automation has a singular purpose: To give machines a chance to perform monotonous, repetitive activities or as specific individuals like to say, "to take the robot out a human." It saves time for individuals to concentrate on increasingly significant, innovative tasks that require human touch and judgment. The final product is an increasingly proficient, practical business, and a progressively productive workforce. A submissive advanced robot that never calls in sick or takes days off, and consistently takes care of business. It is no big surprise that organizations so readily adopt automation.

Automated machines are driven by the manual configuration such as work processes, programming edge case situations, and the like. Basically, a soldier machine that follows orders.

Automation will progressively affect the industrial revolution during the following couple of years. The effect of automation is already felt in particular businesses, and some nations are encountering the impacts more so than others.

The specialized possibilities for automation differ significantly across industries and activities. As automation innovations, for example, robotics and machine learning assume a tremendously incredible job in regular daily activities, their likely impact on the working environment has become a focal point of research and open concern. The conversation inclines toward a Manichean speculating game, which asks whether machines will succeed jobs or not.

While automation technology will dispose of few occupations in the following years, it will also influence parts of practically all employees to a higher or lower level, depending upon the sort of work they involve. Automation, presently progressing past standard manufacturing, has the likelihood, at least as to its technical feasibility, to change sectors, for example, medical and financial, which include a significant portion of information duties.

Understanding Automation Potential

In talking about automated technology, we allude to the possibility that a given task could be automated by adapting current tested innovations. In other words, regardless of whether the automation of any device or activity is feasible, every occupation is comprised of different sorts of duties and tasks, each with fluctuating feasibility levels. Occupations in retailing, for instance, include activities like gathering or handling information, interacting, and relating with clients and setting up product display.

Technical feasibility is an essential prerequisite for automation; however, not a sole indicator of whether an activity may be automated. A subsequent factor to assess is the expense of creating and sending both the equipment and the product to be automated. The cost related supply-and-demand and labor elements speak to another ensuing factor: if perhaps laborers are many and mostly more affordable than automation, this eventuality could be a decisive contention against it. An additional factor to examine is the advantages past labor-cost exchange; including more elevated amounts of yield, higher quality, and fewer mistakes. Administrative or social issues, for example, the value of machines in a particular setting, should likewise be gauged. A robot may have the option to replace some activities of a medical physician, for instance. But for the time being, the prospect that this may actually occur could prove disagreeable for some patients, who are accustomed to expect human contact.

The likelihood for automation to sprout in an industry or occupation mirrors an inconspicuous exchange between said activities and interchanges among them.

Even when machines eventually assume control over several human duties in a certain occupation, it does not quite spell doom for the job in that profession. Despite what might be expected, other professions grow to fill the loss. In the United States, for instance, the massive large-scale distribution of standardized barcode scanners and related point-of-sale frameworks in the 1980s lowered the cost of labor per store by up to 4.8%and the cost of groceries by 1.2%. It additionally empowered various advancements, including increased promotions. In any case, cashiers were also required; indeed, their work demand grew at an average rate of 2 percent between 1980 and 2013.

Partnership Between Machines and Humans

The greatest minds of this generation, Tesla's Elon Musk and Stephen Hawking, disapprove of artificial intelligence in a manner that certainly falls on the dystopian saying that "Robots and AI will have the option to show improvement over us, creating the greatest risk that we face as the human race."

Then we have robust AI advocates who propose that AI will support people but will not control or meddle with their lives. That is the general purpose of AI: To make advances that capably emulate what a human can say, think, and do, which generally would not be influenced by natural delicacy (people age and die).

Also, much the same as most people, that implies AI is terrible at mainly following orders. That is not what it is intended to do; it is meant to always look for patterns (like people), gain from experience (like people), and self-select the suitable reactions in the circumstances (like people).

In this way, AI is not a replica of me or you. It is tied in with making a system that is more powerful than we can envision.

What drives both automated systems and AI is something very similar that drives organizations: data. Businesses that are automated perform better and gain significant revenue growth. This, obviously, might be an aftermath of numerous variables, including the standard cross-segment of advantages related to automation. Expanded profitability; better business effectiveness, and most significant-

representatives ready to concentrate on innovative or/and critical tasks.

We are not yet AI-ready. We are in the wake of cognitive automation. If we continue putting resources into smart automation by fueling it enormous measures of data, we can turn out to be significantly more powerful—as a people, and as organizations.

If we happen to imagine the world in the year 2030, the association among people and machines will reshape lives. Machines will supplement human abilities and help them accomplish more prominent efficiencies.

Future Partnerships

Individuals have lived and worked intimately with technology, or machines, for a considerable length of time - from typewriters and PCs to the expansion of cell phones in our everyday lives. And, owing to massive advancements in software, big data, and processing power, we have entered a world with stunning new conceivable outcomes. We are going to observe an ocean change in our association with machines-portrayed by considerably more productivity, unity, and possibility than before.

As technology power increases 10x in five years, so will our dependence on technology, giving rise to a symbiotic relationship. People will bring abilities, for example, inventiveness and critical thinking, which can be applied against the foundation of human experience and cultural setting. And the machines will bring speed, automation, and radical new efficiencies. By training machines to improve their understanding of people, society, and organization, more of us will readily interact meaningfully with machines.

Profitability will increase, and new businesses and jobs will be made because of this new dynamic partnership. Machines will not replace us; instead, they will enable us to accomplish many more tasks.

People as Digital Conductors

Human-machine joint effort empowers organizations to associate with representatives and clients in a novel, increasingly viable ways. AI assistants like Cortana, for instance, encourage correspondences between individuals or in the interest of individuals, for example, by recording a forum and sending a voice-searchable copy to the individuals who could not grace the meeting. Google and Amazon personal assistants can be integrated into homes, vehicles, and smartphones to do things like order products online, play music, book salon appointments, as well as offer customized fashion advice.

In 2030, we will depend on machines to oversee significantly more parts of our personal lives. We will successfully move toward becoming digital conductors. Technology will work as an expansion of ourselves, helping us better manage our day by day activities. With AI virtual assistants taking care of fundamental requests, human representatives can focus on tending to increasingly complicated issues, particularly those from troubled clients who may require extra support.

Rethinking Business

The ramifications of this partnership stretch well past our personal lives and into how we manage businesses. The work environment will get a makeover as far as how it discovers ability, manages assets, delivers service, and facilitates careers.

By 2030, work will not be a particular place but a progression of tasks, which will be re-appropriated to the best talent the world over. Smart analytics, data visualization, and reputation engines will make people's attitudes and skills accessible, and organizations will seek the best talent and reduce personal bias.

In every way, regardless of whether it is finding the best talent, training employees, or calling upon an entire menu of services, deeper human-machine associations will be the power for change in 2030. We are approaching an era where every business will be a technology-based business controlled by software. Emerging technologies will reshape our lives and work forever.

Learning On-The-Go

By 2030, the capacity to gain new information will be valued higher than the knowledge individuals currently have. Not only will employees have many jobs, but also the tasks and obligations of the occupations will be notably different from what they studied. They will gain the abilities and skills they require to execute their work effectively. They will routinely improvise, learn from one another, and make their way. These variables, combined, will genuinely challenge conventional organizations. Most will band together with machines to learn while 'on-the-gig.'

Human Assisting Machines

People ought to assume three paramount roles. We should train machines to carry out specific tasks; explain the results of completed tasks, mainly if the outcomes are illogical or questionable; and maintain the accountable use of machines (by, for instance, keeping robots from hurting people).

Train

Machine-learning algorithms must be instructed on how to execute the work they are intended to perform. Therefore, massive data sets on training are collected to show machine-interpretation applications to deal with everyday articulations, medical applications to identify disease, and engines built for recommendations to help financial decision-making. Additionally, AI frameworks must be tutored on how best to relate with people. Examples of these already exist like human coaches were expected to build up the characters of Amazon's Alexa not to mention Apple's Siri to guarantee that they precisely mirrored their organizations' brand lines. Siri, for instance, has only a dash of brazenness, as customers may anticipate from the tech giants, Apple.

Simulated intelligence aides are currently under training to show progressively subtle and complex human characteristics, for example, compassion.

Explain

As AIs increasingly arrive at resolutions through opaque procedures (also known as the black-box problem), AIs require human specialists in the discipline to disclose their behavior patterns to non-expert users. These "explainers" are especially significant in proof-based businesses, for example, medicine and law, where a professional needs to see how a machine gauged contributions to, sentencing, or diagnosis, for example.

Explainers are likewise significant in helping insurance providers, and lawyers comprehend why a self-driving vehicle took steps that prompted an accident- or avoided one. Moreover, explainers are becoming a necessity in managed businesses- in any customer-front industry where an AI's output could be tested as unjust, illegal, or biased. As an example, the European Union's most recent General Data Protection Regulation (GDPR) presents customers with the privilege to get clarification for any design-based decision, for example, the service tax offered on a Mastercard or a home loan. Explaining one sector where AI will add to expanding employment. Experts gauge that organizations should make around 75,000 unique openings to cater to the GDPR stipulations.

Sustain

Notwithstanding the people who can clarify AI results by explaining, organizations need "sustainers"- representatives who persistently work to guarantee that AI systems are operating appropriately, securely, and dependably.

AI can help our logical and executive capacities and elevate creativity. Take, for instance, a variety of specialists referred to as safety architects solely focus on preparing for, and preventing harm by AIs. The designers of industrial AIs that work together with individuals have given thorough consideration to guaranteeing that AIs recognize people and not cause them any harm. These specialists may likewise audit reports from explainers when AIs happen to cause harm, as when an autonomous vehicle is associated with a fatal collision.

Other sustainers ensure that machine frameworks maintain moral standards. If an AI framework for credit endorsement, for instance, is observed to discriminate against specific individuals, these ethic managers are in charge of investigating and undertaking the issue. Assuming a related role, data compliance officials attempt to guarantee that the collective data that is fueling AI systems agrees to the GDPR and similar consumer-protection guidelines.

A relevant data application job includes guaranteeing that AIs manage data mindfully. In the same way, as other tech companies, Apple uses AI to gather insights regarding clients while interacting with the organization's gadgets and operating system. The point is to enhance the client experience; however, unregulated data collection would likely compromise security, outrage clients, and cross paths with legal lines. The organization's "differential security group" attempts to ensure that as much as the AI is looking to gain as much as statistically possible, it is protecting individual clients.

Machine Decision-Making

Through giving workers custom information and instruction, AI can enable them to arrive at better choices. Wiser decision-making can be particularly profitable for laborers, where making a better judgment call can profoundly affect the bottom line.

Consider how hardware support is being upgraded with the use of "digital twins"- virtual copies of real equipment. A company called General Electric builds the aforementioned programming models of its turbines and different mechanical items and regularly provides them with updates using the data collected from the equipment. Through gathering reports from many machines out in the field, General Electric has compiled an abundance of data on typical and deviant functionality. GE Predix application, which employs AI calculations, would now be able to anticipate when a particular part of a particular machine may fail.

GE technology has, in a general sense, changed the intensive decision-making procedure of managing industrial equipment. Their application may, for instance, distinguish some surprising rotor mileage in a turbine, check the engine's operational history, report whether the damage has expanded in recent months, and caution that supposing nothing is done, the rotor will be expected to lose on average three-quarters of its useful life span. The framework would then be able to propose suitable moves, considering the equipment's present condition, the working environment, as well as amassed information about similar deterioration and fixes to different equipment. Alongside its suggestions, Predix can produce data about the expenses and money related advantages and give a certainty level of about 95% for the inferences applied in its investigation.

Without Predix, employees would rarely detect the rotor wear-and-tear on a standard maintenance check. It is conceivable that the damage would go undiscovered until the engine malfunctioned, bringing about an excessive closing. With Predix, technicians are cautioned to likely issues long before they manifest, and are provided the required data readily available to use sound judgment that can once in a while spare GE thousands of dollars.

Fundamental Differences Between Human and Artificial Intelligence

Feeling

Human intelligence is powerful because it is not constrained to objective reasoning. Different components of our cognizance empower us to manage the characteristic unpredictable and uncertainty of our general surroundings. They enable us to settle on choices based on shared qualities and inspirations that resonate collectively and empower us to know what is right without having to understand. Human intelligence enables emotions to influence their decisions and their conduct towards others.

A machine cannot do that, regardless of whether it needed to. Things become complicated when machines start making decisions that have significant outcomes, without the emotional setting and shared qualities that all people use when settling on such choices. Machine consciousness is unable to experience the feeling of pain, sweetness, agony, or even the emotions that music or colors elicit.

Creativity

Putting aside whether human creativity is limited and, what exactly creativity is, it is unquestionably evident that artificial neural systems being grown today work out the rules as they come, instead of being taught. AlphaGo, the AI that crushed the Korean go grandmaster Lee Sedol, was fed a large number of games, however no rules. It worked out how to play go without anyone else's input.

In any case, can machines indeed be creative? Would they be able to be viewed as artists in their own right? If creativity characterizes being human, by what method can a collection of wires and transistors be considered to be imaginative? Maybe we are not as different as we might think. Our human minds are too constrained to even consider imagining how incredible machine imagination may become. Today's machines already display glimpses of creativity in art. There is software that generates convincing images of non-existent people by piecing together algorithms of facial structures.

Machine Thinking to a Human Problem

At the point when there is no data, there is no AI. In that capacity, AI cannot solve what has not been collected in terms of data. Therefore, AI has a limit of knowledge.

Unless data has been fed to the AI system such as detecting emotions, heart rate, facial expressions, movement, routine, and the like, the system would not be able to understand human behavior. It will not adapt to different sets of motivations to which a human being is wired.

The unmistakable fixing in human knowledge will be viewed as the ability to incorporate informed and passionate speculation to make moral choices are adjusted to the unique circumstance. It uncertain that animals can coordinate, all in all, objective reasoning, and emotional deduction to make moral choices. Regardless of whether a few animals might do it, the suspicion here is that the sort of ethical quality rose would be extraordinary and reasonable for every species, and culture.

Consciousness

By the age of four, children ordinarily start to get a handle on one of the essential standards in society: that their minds are different from other minds. They may have different convictions, wants, feelings, and goals.

Machine consciousness is another, mind-boggling and interdisciplinary research zone. It is firmly interlaced with AI, and even though it would be simpler attempting to disregard this point, it is a fundamental issue for everybody who needs to approach AI genuinely and from a foundational viewpoint.

Although there is immense interest in artificial intelligence, there has been significantly less interest in the collective consciousness. That is one motivation behind why we have additionally seen little improvement in how smartly our governing systems work- democracy system and governmental issues, business, and the economy. We are encompassed by organizations pressed with human intelligence that all things considered frequently show collective consciousness.

Even though there are no advancements in the area of machine awareness, or it is demonstrated that machine cognizance does not exist, it will add to the learning and systems of human consciousness. This will ideally prompt precious advancements in the fields of health, administration, and politics.

All analysts researching machine awareness will be faced with legal and moral questions and suggestions at some point or another. This can occur in different situations, similar to the animal protection movement has gone up against research and practice as a result of the acknowledgment of animal awareness.

Uniqueness

The similarities and differences do not imply that AI is better than human intelligence, or the other way around. The point is, they are very different things.

Artificial intelligence is great at redundant tasks that have explicitly characterized limits and can be spoken to by data and awful at expansive errands that require instinct and decision making dependent on incomplete information. Interestingly, human knowledge is useful for settings where you need the presence of mind and conceptual choices and terrible at tasks that require substantial calculations and real-time data processing. Human intelligence and AI supplement one another, making up for one another's inadequacies. Together, they can perform undertakings that none of them could have done exclusively.

As AI gets better at performing an ever-increasing number of tasks, we as humans will find more opportunity to put our intelligence to genuine use, at being creative, being social, at literature, poetry, sports, and every one of the things that are important on the grounds that the human component and character that goes into them. Furthermore, we will utilize our expanded insight devices to improve those creations. The future will be one where artificial and human intelligence will build together, not apart.

Chapter 3: The Technology Behind Human Machine Interface

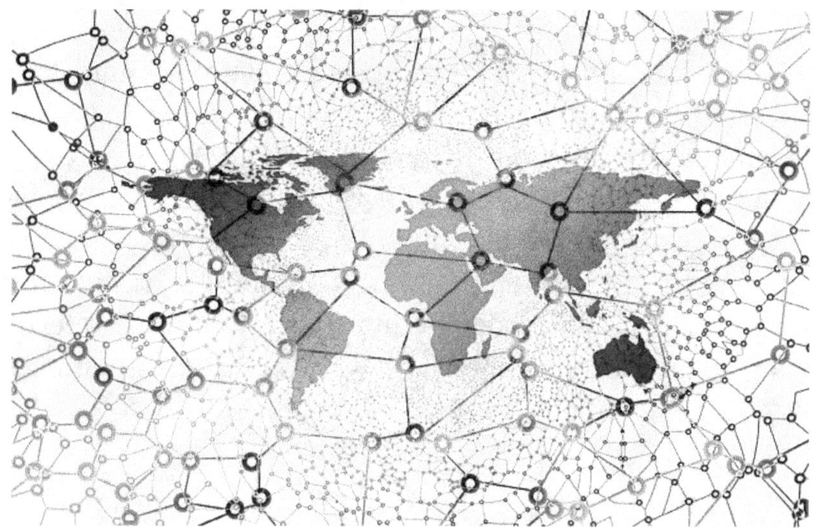

"Technology is changing the way we interact as humans" —
Dan Brown

In this chapter, we will cover the following:

- Definition of Human Machine Interface(HMI) and its uses.
- Styles of HMI.
- Anthropoid Machine Interface.
- Smart automation influence in lifestyle and relationship
- Hard and software.
- Configuration type.
- Market dynamics.

Definition of Human Machine Interface (HMI)

The Human-Machine Interface (HMI) is a constituent of precise expedients that are skillful of controlling human-machine relationships. The engine boundary consists of hardware and software that will permit operator inputs to be converted as indications for machinery that will again give out the mandatory outcome to the operator. HMI expertise is being cast-off in countless productions like microelectronics, showbiz, and medicinal, military amongst others. It has helped in the integration of humans into multipart high-tech systems. The human-machine interface can also be referred to as man-machine interface (MMI), human-computer interface, or computer-human interface.

HMI has two interaction types, i.e., humanoid to engine and engine to humanoid. Being that HMI is ever-present, boundaries tangled can consist of wave radars, controls and marginal expedients, dialog acknowledgment, interaction info exchange using light, sound, high temperature, and other perceptive means that can be considered part of HMIs. HMI can also be used as an adapter for other technologies despite being considered as a standalone technological area. They are built based on humanoid bodily, behavioral, and psychological competences, which means that ergonomics will formulate ideologies behind HMIs. They can be responsible for matchless chances for submissions, wisdom, and restoration, apart from being able to enhance user experience and efficiency. Better HMI can give excellent and natural relations with exterior devices. The diagram below illustrates the HMI.

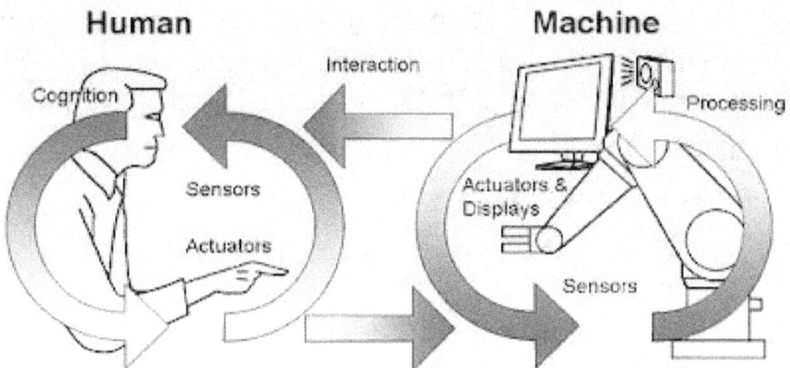

Figure 1: A HIM interface

There are advantages given by incorporating HMIs included in the inaccuracy discount, amplified scheme, and consumer effectiveness, upgraded consistency, and maintainability amplified consumer approval and consumer wellbeing, decrease in preparation and expertise necessities, bodily users, duty overload, amplified budget industrialized, and efficiency among others. Touchscreens and casing modifications can be referred to as samples of HMIs. Its knowledge is castoff largely in key and flat performances, array response, internet entrance, records involvement for microelectronics, and info synthesis. There are certified groups like GEIA and ISO provide standards and guidelines that can be applied for Human-machine interface technology.

Uses of Human-Machine Interface

HMI system:

  • HMI is a software that has been designed for the interface between the user and the machine.

  • HMI is a device that acts as a border amid the PLC and the operator.

• HMI is used in industries to enhance communication between machinery and production plants.

• HMI will provide a graphics-based visualization of industry control and monitor its system providing real-time data procurement.

HMI Elements:

The elements consist of:

- Development of static screen.
- You are linking of dynamic.
- Trends
- Setting up alarms and securities.
- HMI and interactions amid PLC and HMI.
- Advantages of HMI
- It's easy to use for non-technical individuals.
- It's user-friendly.
- Recording of data.
- Enhanced productivity.
- Enhanced communication.
- Reduced costs.
- Easy management of the plant.

Human-machine interface goods feature the essential electronics to regulate and signal different types of automation apparatus in an industrial setting. HMI goods range from simple designs with standard LED indicator screens too much more multifaceted HMI systems with touch screens and many features. HMI systems must be vigorous and be capable of enduring the harsh environment. The HMI should be resilient to dust, water, and extreme temperatures and even at some point, harsh chemicals.

There are very many uses of installing HMI in your plant or facility as discussed below:

Alarms

Plant operators can see signals that have been given by the human-machine interface to identify any malfunction of equipment and react quickly. Alarms help in preventing and alerting the operator before an emergency level is reached. Alarms can also help in tracking other problems and help in optimizing production processes increasing productivity.

Reliable Messaging

As an operator, you can depend on HMI messaging for Facebook pages, faxes, and many more when there is a particular event. On other occasions. Operators can be alerted automatically when fuel levels of a machine are low, and there is a need to refill.

Easier General Management of Plant

HMI technology will efficiently manage and execute recipes. It has high-quality graphics that provide a realistic view of the plant operation, giving the operator easy control from a central monitor.

Testing Accurately with Simulation

Plant directors can check devices and equipment quickly by the use of a flexible HMI with simulation. You can achieve the testing from the comfort of your office without any equipment. This will help in reducing the starting time and enhance overall production.

Reduction in Costs

HMI reduces operation costs by replacing hundreds of selectors, indicator lights, and push buttons among many others. Meaning it will reduce the addition of panels, cables, and consoles.

Enhanced Communication

HMI enhances communication between various types of equipment throughout the facility with the use of remote, Ethernet, data highway plus, serial port, dynamic data exchange, and many more methods. You can note that having HMI can benefit your facility in operating, safety, and production. Industries like Red Lion gives a variety of HMI goods even for demanding industrial applications.

Types Of Human-Machine Interface

There are different types of HMI, as discussed below:

Automatic Replacer

It rationalizes the fabrication process by consolidating all the roles of each switch into one position. It replaces LEDs, on/off knobs, and buttons. Abolition of these motorized devices is probable because HMI can offer an optical design of all devices attainable on the LCD display while resounding out all the other tasks.

Handling of Data

This kind of HMI is usually impeccable for applications that necessitate a persistent response from the scheme. They mostly come fitted out with substantial competence memoirs.

You must certify that the HMI screen is enormous enough for diagrams, fabrication breakdown, and optical diagrams. It can comprise roles like guidelines, sorting of data, and fright management.

Administrator

Chief HMI will be important in any presentation that involves SCADA. HMI will utmost, run windows and partake a lot of Ethernet ports.

Selecting of Program Writing Software

At hand are three sets that you can pick out from:

Trademarked Software

This is the software the builder makes available, benevolent it stress-free and permits it to have fast development.

Hardware Self-Governing Software

This is third-party software that has been conventional to suite on numerous varieties of HMIs. It delivers the producer with liberty for HMI assortment.

Vulnerable Software

This software permits the architect to have wide-ranging honesty in the strategy course and only be carefully chosen by a liberal computer operator.

Human Machine Interface Applications

There are various solicitations of the human-machine interface:

- Convenience commerce can practice HMIs to notice water transmission and wastewater handling.
- They are used in the bottling process, where it assists in controlling all the features of the fabrication line like haste, fault alteration, and competence.
- HMI can control the cutting of metals and how faster it is done in metal manufacturing industries.

Human-Machine Interface Advantages

- HMI reduces production costs.
- It improves efficiency and productivity.
- It saves on time, thus increasing profit margins.
- They can save data, which is convenient to be used far ahead for troubleshooting upcoming motorized hitches.
- HMI expedients have extraordinary innovation and capability to elaborate utilities than ever before. They can give high-tech urgencies like disregarding the need for a mouse and controls and talk of hardware to software.

Human Machine Communication

In case you want to be useful, inspiring, and exciting, then you have to create a genuinely cognitive experience. Cognition is a mental process where you acquire knowledge and understanding through thinking, experience, and senses. Cognition will define how human process thoughts to interact with each other. In the case of computers, the perception will refer to a system that arouses human thought processes by the use of algorithmic models that are intended to augment human cognitive skills.

This raises an essential question like: how does the cognitive system arouse human processes? It is because it must activate your human senses; it must create a motivating experience and inspire your thought processes with the same goal of acquiring knowledge and shaping your understanding. This is known as *presence*. There are persuasive cognitive clarifications that create a presence in our lives through that presence, augment of your cognitive abilities.

Compared to non-cognitive systems, cognitive applications will go beyond what you experience today in transactional applications like pushing a button, getting a measured response. IBM can distinguish them as systems that can understand, reason, learn and interact as expected. To achieve this, cognitive systems will analyze vast quantities of data to constitute insightful, contextually aware, and continuously improve connection with users. Due to its growing knowledge of users' needs, goals and values allows them to deliver personalized replies, suggest relevant perceptions, and revealing contextually substantial discoveries.

For you to understand, reason, learn, and interact, there are various elements of human thinking and communication that cognitive systems should be able to recognize, understand, simulate, and analyze:

- Perception
- Motivation
- Learning
- Reasoning

- Knowledge

There are a lot of various levels of cognitive functions depending on the autonomy of the app. The low-level cognitive app requires a lot of support from the programmer, while the high-level app will behave more on its own. New applications might begin with low cognitive function because a user must train knowledge and behavior until the app answers reliably. With time it will develop more innovative cognitive functions. Every cognitive capability can have diverse levels of tasks depending on the quantity of user intervention.

An interaction amid the user and cognitive application might not need every one of these abilities, but the app itself will require the ability to finish a whole cognitive experience something which has a presence with the programmer. Human to machine interaction model ties the components necessary for cognitive systems collected into a methodology for creating cognitive skills. The main resolution is to direct and inspire intentional innovation and give structure for manufacturing responsible design decisions basing on human wants, expectations, and values.

Part 1. Input: Understanding the World

Knowledge

Man's impulse to communicate with technology; hammer, microwave, or quantum computer, will align right with the technology's ability to increase human lives to prolong your strength or reach. In the case of cognitive computing, enhancing that pulls you to interact with the ability to process vast amounts of data that augments your thinking. This will enable you to make the right decisions and bring about discoveries quickly than humanly possible. Due to this unique ability, it helps doctors spend less time to research and create more time to care for patients, creating target lessons for every student's individual wants, assisting industries in serving millions of buyers concurrently, individually, and proactively.

Knowledge can be termed as a summation of the whole thing a cognitive system recognizes from the ground −truth data that it's formerly educated with to learnings of every communication it experiences. Cognitive systems can be trained on any subject and are given a model for that area. They are usually better at reading, identifying, and remembering large amounts of unstructured info in ways impossible for the human brain to process. They can quickly analyze lots of pages of content and give a summary of highlights, or listen to many hours of music then create their songs revealing relationships and designs across formerly independent research. They will improve their capability to give you personalized answers, relevant insights, and discoveries with a new piece of data adding to their knowledge base.

Knowledge can be referred to as the applications ground truth and ever-expanding skill and expertise set:

Good – the form contains subject matter expert knowledge that sorts the application a competent tool for problem answering.

Better – the application will allow the programmer to update the knowledge base with trained data.

Best – the application will update its knowledge base by itself using live sources.

Perception and Motivation

Responding to input from the outside world, cognitive applications will require to understand context, circumstances surrounding an event, statement, or an idea and this is because the system will fully understand the meaning of a programmer's intention at the moment of communication and give you insightful, timely, and natural answers. You will be able to recognize dates, author, and quality of information and validation of sources in regards to incoming articles allowing the cognitive system to tell what priority to offer to the new info. With a negative press regarding a company's good, the CFO will want to start analyzing possible repercussions in stock charges immediately. A cognitive system can couple news alerts with significant insights and stock analysis, knowing it's an urgency for the CFO. The context will come from any source that will distress the system's capacity to offer intellectual answers to a programmer and can be deliberated from two perspectives that relate to the human thought process.

Perception

If a cognitive system realizes that the user is at home and not in the car, it can conclude you are likely inquiring about your dog. Furthermore, if the system recognizes again that the vet recommended you take your dog outside every two hours due to the instructions sent to your mail. Perception is the application's capability to ingest, organize, and classify info about the programmers' physical and digital, recent, and historical context. Perceptual statistics are things such as location, time, date, expression, mood, environment, networks, nearby places, and connected applications. Perception uses APIs to stream information about the world where it includes weather, delays, social media, events, and traffic jams. When there is more data, a cognitive system can collect nearby awareness both historically, and at the moment, more insightful and natural responses can be.

The perception which is the application's capability to consume, classify, and organize information about the programmers' bodily and digital context:

• **Good** – the application will sort and organize information considering its pre-training.

• **Better** – the application is capable to categorize and organize recent information from live sources and what it has learned.

• **Best** – perception application will infer data basing on the other information, i.e., if Brian is in hospital and the doctors instruct him not to take fluids, the cognitive system will realize that he has a glass and he is drinking from it and can pass the info to the doctors that he is hydrated.

Motivation

Accepting motivation will give the cognitive application the awareness about programmers' goals, values, and priorities so that it can modify an insightful answer that will meet the users' anticipations for communication with the system. There is data that will define a programmer's motivation and can be sought out by their setup experience, responses, expressions, preferences, and communications with time. With the growth of programmers' communication history, the understanding of the system of the users' wants, and behaviors also grow improving the knowledge with each communication.

Motivation allows the system to understand and prioritize behavioral and personal info about the programmer's reasoning to come up with a valuable response. This is done by evaluation of the success of previous reactions when the user was in the same circumstance that defined your wants and values at that moment i.e., a cognitive system can decide not to interject a call with a work notice because in the past the programmer lay off similar notice when on the phone with the mum. It can decide to alert a programmer to a news feed you aren't pledged to because the user currently is concentrating on this current topic at work. It can also listen in a meeting and resolve to send the team response on refining how to operate stand-ups in line with minutes of the meeting, considering the managers' goal to enhance responsive practices.

Cognitive systems are required to account for why a programmer is inquiring about a particular order. Consider knowing the expectations for communicating with a system that is built by an athletic clothing trademark against a system developed by a music review trademark. In case a programmer inquired every order on what to do on a Saturday night, you would expect a range of answers basing on the values the trademark of that system symbolizes. Cognitive systems have objectives and values that are demarcated by their inventors that need to be accounted for and conveyed in their reactions to meet the programmers' anticipation.

Take a situation where perception and motivation are the significant components for making a programmer feel understood. The system is expected to reflect ways that will the programmer is well informed of and can remember past interactions, anticipates wants without any directions. It should be able to reduce friction and cut the number of steps it will have to finish a task. If it's done in the right way, then the cognitive system will have a feeling that knows you and understand your wants.

Motivation, well known as the applications ability to understand the programmers' intention, priority, values, and goals:

•   **Good** – the application is well conversant with demographic factors and trade focuses of its programmers and surface info for that reason.

• **Better** – the application can identify individuals and their personalized characteristics. It can be familiar with users' feelings and answer back with the most applicable emotion.

• **Best** – the application proactively communicates with the programmer basing on how you will answer back, i.e., an assistant is well conversant with the person to assist and plans on how to cater to their wants.

Part 2. Output: Natural Response

Reasoning

Applications mostly have intelligent communications with a programmer by easily giving the ground truth info that is stored in its knowledge base. By the use of its experience about the context of a user, the cognitive application will go beyond accurate translation and retort with a more treasured, big-picture answer. Reasoning can be referred to as the application's ability to have cognitive communications by considering all the info available through perception, knowledge, and motivation. If a solid knowledge base produces an intelligent response, it won't feel cognitive if the application doesn't serve the individual and consider your context in some way.

By having confidence scores to potential answers based on detailed findings and previous communications, the system can think about how to make up a comeback that is modified and foretelling. Single communications can have more abilities in use at the akin time, but the cognitive application will compose its response based on what it's studied to date, and aim at improving the answer in the future based on the programmers' reaction to that particular response. The reasoning is an application capable of having intellectual communications basing on contextual and historical knowledge of the programmer:

• **Good** – the app can generate predetermined responses that are precise to the domain or directed problem space. It necessarily doesn't use perception or motivation abilities when creating responses but relies on pre-trained knowledge.
• **Better** – it creates resourceful responses and relies on perception, motivation, and knowledge abilities when creating responses.
• **Best** – it will bring anticipation to the user's want, and respond directly to them creating recommendations that will have a benefit to your needs and context far much ahead anything stated explicitly.

Learning

Where there is an interaction, cognitive applications will update their knowledge about a programmer, current data, and the world based on the programmers' response. There are situations where the user immediately hit off the link that was suggested without checking on the content. Cognitive systems usually update the way they interact with people basing on findings from personal and collective historical experience. They remember past communications and adjust responses based on those studies by enhancing adjustments to the confidence scoring of context in the matrix. Learning considered as the applications capability to interpret user responses and apply that knowledge to improving communications over time:

- **Good** – it will allow the programmer to train pre-packaged info in the interface. It doesn't teach the perception or motivation abilities and its purpose to create more trained knowledge.
- **Better** – application learns through programmer communication and behavior and explicit feedback. It trains perception and motivation abilities.
- **Best** – the application updates or trains its info without any user intervention.

Smart Automation Influence In Lifestyle And Relationship

Previous years there have been major interruptions in commerce models. The maladies of the bygone are now megatrends for the forthcoming, and emerging technologies and digitization are guiding many of them. Many buyers have moved from purchasing goods to consume services, experience, trades having no choice but to change round their operative models fundamentally. You can achieve this by automation of supply chains. There are different ways to foresee this coming in practice above the following five years:

1. Robots Fitting Together the Industrial Sector

In built-up company's, robotics, and artificial intelligence (AI) will be approved in the next five years. It can be seen happening already, i.e., rather than having a contract out built-up to Asia, Zara fabricated 14 highly computerized productions in Spain having robots working by cutting and dyeing textiles thus permitting the commerce to embrace time-saving catalog tactic.

Customer markets experiencing rapid growth will want to increase production abilities. From 2013 china has been graded as the world's chief market for robotic technology. Upturn in the use of robots and AI can generate first-hand hope for countries like China that are trailing significance effectiveness in case of rapid rising factory bills and the increasing challenge of getting production workforces.

2. Massive Data and Analytics Will Be Castoff Across Resource Chains

Cost of sensors and microprocessors making production smart retain going down, motivating the comprehensive solicitation of engineering automation. Some industries cannot substantiate the worth of predictable systems, veil computing, and software as a provision, where software is centrally accommodated making it obtainable on a payment basis making the statistics organization and analytics easily manageable. Using big figures and analytics with the radar, robotics and AI technologies will together produce a smart end product.

3. Last-Mile Delivery Will Be Uberized

When it comes to distribution, companies are already working out on how to solve the last mile problem, i.e., people are allowed to edict from indigenous grocery online and refer a private purchaser to pick the kinds of stuff and supply. Other corporations are working on solutions like a Uber rush to get to the clients at the exact time. Others are at work on improving the superiority of facilities and broadening the structural analysis. In the near-term five years, a lot of similar ideas would be there to fit anyone's needs.

4. Expiration of the Great Street

If purchasers are not buying online, they can be over and done in the next five years. Attempt to deliver goods to buyers as customers faster, industries like the amazon will create more regional delivery centers. There is a Chinese who have invested in constructing its supply network. It has 82 storerooms with a total area of more than 1.3 million square meters in 34 municipalities across China with 1453 local delivery stations in 460 cities.

Not only trades that feel the waves, in the succeeding five years, but there will also be a renovation of the fund chain which will bring influence to people in countless substantial ways:

Rising of Industrialist—Above the past two eras, the growth of corporations such as eBay has conveyed up a fresh form of micro-entrepreneur. Most current scientific improvements will improve even additional methods for persons to earn a living, i.e., Etsy an online bazaar vending with arts and crafts that allow hobbyists ton to retail their merchandise globally. This shows that although supply chain automation modifications the world of toil, it will devise to offer countless current models of occupation.

Putting Intelligence to Use—Regardless of the high technical improvement, commercial is still considered to meet people face to face, interact, and build a relationship. The systematic transactional side of the vocation can be robotic and digitized; a lot of trade enlargement is all about devising first connections and offline management. New skills will, consequently, be well-thought-out to upturn associated to the heights they are at today.

Hardware And Software

Software: it is a universal word castoff to define a gathering of CPU programs, processes, and credentials that implement some task on a CPU system. CPU systems will distribute the software organizations into three prime kinds: system software, package software, and solicitation software. It can be stereotypically automated with a user-friendly boundary that permits humans to interconnect more competently with a CPU system.

Hardware: It can be pronounced as a device like a hard drive that is obviously or bodily associated with a PC. Examples of a hard disk are CD-ROM, PC monitor, copier, and film card. A computer is bound not to function without hardware making it difficult for the software since it has nothing to operate. Both soft and hardware communicate with each other where software tells the hardware which duty it has to perform.

Contents of Hardware and Software

- Software
- Firewalls
- Function
- Changes
- Interdependence
- References

Type

Hardware is a physical device you can touch and see like a computer monitor and software programs can instruct the equipment to do a particular duty. Comparing the two, the software has no physical form. Both of them are associated with computers, and there are situations where software runs hardware like phones, GPS, medical equipment, and air traffic control system. Without software programs, machines would be considered useless, i.e., you can't communicate with the computer when it doesn't have a software operating system.

Function

The software executes a specific task by handing out an ordered set of programmatic instructions to hardware. Hardware will serve as the delivery system for software clarifications.

Interdependence

Hardware can't operate until the software is loaded and installed in the hardware to set the programs in action.

Firewalls

They are available for hardware and software. Popular firewall choice is a software firewall that is installed on the computer and can be customized to suit a personals user security wants. They are usually found in broadband routers.

Changes

It has been common to switch to new software or to use several kinds of software at a time; hardware is usually less changed. Software is created quickly, modified, deleted, but hardware needs a lot of skills and can be an expensive endeavor.

Configuration Types

Besides exporting a single configuration object, other ways cover the needs as well.

Extending Configuration Types

Extending an existing configuration type, you need to come up with a configuration file in your module, i.e., add an event viewer you create, *app/code/{vendorname}/etc/events.xml* and declare a new viewer. You can find out that the configuration type is in existence in Magento, the loader and the functions authenticating schema by now present and functional.

Creation of Configuration Types

For the creation of a new configuration type, you at least have to add the following:

- XML configuration files

- XSD validation schema
- A loader

To introduce an adapter for a new search service that enables extensions to configure its entities, then you have to create:

- Loader
- XSD schema
- Additional classes for your new type of work
- Appropriately named the configuration file

Exporting Functions

You can find the need to disambiguate in your webpack.config.js amid development and manufacturing builds. You can export a function from your webpack config as an alternative to exporting an object. The function involves two arguments:

- Environment as the first parameter.
- An options map as the second parameter.

Exporting Promise

Webpack can run the function exported by the configuration file and wait for a commitment to be reverted. It can be handy when you need asynchronously load configuration variables. As an alternative to exporting a single configuration function, you may export various configurations. All configurations are built.

Market Dynamics

It is referred to as the factor which distresses the supply and demand of products in a market and is essential to economics as they are to practical business application. A lot of economists establish market dynamics. They have been considered to be the most developed in Porter's Five Forces of competition.

Market dynamics can also mean elements that affect the market. From the study of economics, they supply, demand, price, quantity, and other precise terms. From a trade standpoint, market dynamics are factors that affect business models involving the applying party. When compared, the dynamics may be the price of a barrel of crude, oil manufacturing, national stockpile, and more for oil firms.

For a sensible trade, market dynamics is incorporated in the market analysis of their business plan. These factors have affect trade so much that it would be neglected if you don't exclude them. Market dynamics play a vital role in the marketing plan of a business. They can play a critical role in various areas like the cost of products sold, distribution, logistics, and many more.

Examples of the Market Dynamic

Beryl is writing a market plan for her new trade. She sees a real want in the fashion company for high-quality accessories like purses and necklaces. She will have a great experience in retail, giving her a strong base to refer back to. For her to finish her marketing plan, she has to complete competitive and industry analysis. She will have to assemble the market dynamics analysis for the fashion industry respecting accessories.

Beryl will also need to understand them to fill space for buyers not being served.

She will have to start with the company analysis. Looks at various statistics: buyer spending rate, retailing growth, fashion growth industry, brick growth, and mortar trade sales, and competencies of retailers. Beryl is collecting market dynamics to know whether the market can upkeep the business. She is sure whatever is done is a fundamental foundation for her hint.

Market dynamics come as a result of collective market resources and preferences. Due to this reason, market dynamics are unaffected by the actions of any individual or industry. Individuals act in response to market dynamics as an alternative to causing them. The market dynamics are active in health, organic, natural, and eco companies over their market LOHAS division. A founder of Mambo Sprouts Marketing Market Dynamic is a company leader in publicizing and promoting services reaching health consumers online at wholesale and by emails.

Market Dynamic Services

- Market analysis and targeting publicizing
- Brand research and promotion of evaluation
- Demographic statistics analysis, census, and bureau of labor
- Marketing research, communal mapping and wants assessment(non-profit and trademark)
- Third-party industry/trademarked research and promotion amenities
- Marketing communication, direct mail, e-marketing, and PR
- Company leadership assessments
- Business marketing research and insights
- Communal partnership and local trade visibility
- Mobile trademark, advertising, and promotions

In financial marketplaces, some and not all, financial services specialists are knowledgeable about how markets work. These specialists make rational resolutions that are in the best interests of their customers basing on all the available info. Savvy specialists are basing on extensive analysis of extended knowledge, and proven methods. They work to sufficiently bring understanding to their clients' wants, goals, time horizons, and capability to withstand investment risks.

Chapter 4: Artificial Intelligence Key Leading Players and Promising Startups

Artificial intelligence is moving forward very fast. It is now playing a critical role in your technological world across all industries, from Agriculture to Communication. From Google to Apple and Microsoft, every major tech company is dedicating a lot of resources to breakthroughs in artificial intelligence. Every company now knows you must-have Machine Learning in your story to stay competitive. Several private companies wield immense power and influence in the world of A.I. both in the U.S. and across the globe. Such companies are using the technology in a meaningful way, and they are succeeding in demonstrating real business potential from doing so. Here is a list of the big players which have power and resources to shape your future using artificial intelligence technology:

Twitter

Twitter has invested a lot of resources to get into Artificial Intelligence Learning. The company has already acquired four artificial intelligence companies in quick succession. The most notable of this is the Magic Pony. The company acquired Magic Pony for a whopping $150 million. The A.I. company has developed Machine Learning approaches for visual processing on the web and mobile devices, and Twitter will use this to improve its systems for recommending specific tweets in your timeline in the near future.

SenseTime

SenseTime artificial intelligence company already has big clients, including the Chinese government on its list. The company is supplying the Chinese government with face-recognizing technology to track her citizen. The technology has been lauded as the best, better than what Google and Facebook are doing when it comes to face recognition. Currently, SenseTime is concentrating its resources on developing autonomous driving technology.

Qualcomm

Qualcomm is a chip manufacturing company that is committed to artificial intelligence. A.I. plays an essential role in 855 mobile platforms developed by the company. The chip uses a signal processor for AI speech, audio, and image function. Qualcomm Snapdragons powers some of your most popular smartphones in the market today. Keep a close eye on Qualcomm if you are a fan of the smartphone technology for new exciting AI smartphone features.

Nvidia

Nvidia is perhaps one of the longest established AI companies and still plays a crucial role today. Nvidia's graphics are the be-all and end-all for machine learning and artificial intelligence. The company is active in Healthcare, higher education, retail, and robotics. The company is developing AI technologies that will be integrated into every level of vehicle manufacturing and autonomous driving.

Microsoft

Microsoft is involved in Artificial intelligence on both the consumer and business side. Cortana, Microsoft's AI, digital assistant, is in direct competition with Amazon's Alexa, Siri, and Google's digital assistant.

The Artificial intelligence features use the company's Azure Cloud service to provide chatbots and machine learning to some of the biggest names in the business. Microsoft has also purchased several AI companies in 2018 alone.

Intel

Intel has been on a shopping spree for several artificial intelligence companies. Intel has acquired Nervana and Movidius and several smaller A.I. startups in the recent past. Nervana enables companies to develop specific deep learning software while Movidius was founded to bring AI application devices with deficient performance. Intel has also partnered with Microsoft to provide AI Acceleration for the Bing search engine.

IBM

The company has been active in AI ever since the 1950s. It is one of the pioneer companies of artificial intelligence and is still active up to date. With Watson, IBM has come up with a machine learning platform that can integrate artificial intelligence into business processes such as building a chatbot for customer support. IBM's clients include big companies like KPMG, Four Auditor and Brazil's bank, Bradesco.

HiSilicon

HiSilicon is the manufacturer of Huawei's Kirin 980 chip which was unveiled at IFA 2018 in Berlin. Karin 980 has significantly enhanced the second generation of the world's first Artificial Intelligence smartphone chip. The chip can do things like face recognition, intelligence translation, and image segmentation at a very high speed. The chip has led to the development of many other smartphone chips by the competitors.

Google

This is one of the largest and most important AI company. Over the years, Google has been acquiring several AI startups at crazy speeds. It has also created over 12 new artificial intelligence companies in the recent past. Google acquired DeepMind for a sum of $400 million. DeepMind is the board game playing champion.

The company is also funding the ongoing Tensor AI chip project for Machine Learning on the device. The company hopes to evolve from a mobile-first to an AI-first world in the computer industry.

Facebook

Facebook has allocated immense resources in artificial intelligence, perhaps in recognition of how AI technology will play a critical role in the future world of business. Facebook's AI research group is known as FAIR, has been hard at work to advance the field of Machine Intelligence and in developing new technologies to provide people with better ways to communicate. The company also worked with two AIs known as Alice and Bob among others, but the project was terminated prematurely because of a technical malfunctioning that enables the couple to communicate in their secret language.

DJI

This is perhaps one of the most famous Chinese artificial intelligence company. The company is currently valued at 15 billion dollars, with a 70 percent market share in the global drone market. The company has been increasingly entering the A.I. market with the latest drones using A.I. and image recognition to avoid crashing into objects while on the flight. The company is also looking at entering into the lucrative autonomous vehicles and robotics technological markets. DJI has also recently partnered with Microsoft for a drone to computer streaming project.

Banjo

The company was started soon after the Boston Marathon bombings in 2013. The company uses AI to search social media to identify real-time events and situations that can be critical for emergency services and other organizations to operate faster and smarter. The company has attracted tremendous investment partners, including SoftBank, the Japanese telecommunication giant.

Apple

Apple sees artificial intelligence has a critical part of its future. Consequently, the company has been busy acquiring A.I. startups in recent years. The company has developed products such as Siri and the company's newly Create ML Tool, which macOS and iOS developers to create efficient and straightforward training courses for their apps.

Viz.ai

The company aims at reducing the number of stroke victims who don't receive the right treatment in time. Its software cross-references CT images of a patient's brain with its database of scans and can alert specialists in minutes to early signs of broad vessel occlusion strokes that may have otherwise been missed or may have taken too long to spot. The AI product from the company is already in use in some major hospitals in several countries such as Mount Sinai in New York and the Swedish Health System in Denver.

Deep 6

The company has developed an AI product that helps pharmaceutical research teams to find the right cohort of patients to work with on their new technological trials. The company's software can pull data from electronic medical records to create patient graphs that allow researchers to filter for specific conditions and traits, leading to matches in short periods of time. The system's language understanding engine has been trained to infer some situations even if they are not mentioned in the notes. Currently, the company has more than 20 health systems and pharmaceutical clients in the states and across the globe.

Armorblox

Armorblox launched into the cybersecurity market two years ago. It aims to protect customers from socially engineered attacks, like phishing emails which takes advantage of your missteps. It uses natural language processing which allows machines to learn and understand the language. Its software analyzes a customer's communication styles to get a sense context and then automatically flags possible phishing attempts, insider threats or accidental data disclosures.

DefinedCrowd

DefinedCrowd taps human contributors to build bespoke datasets for clients. The company recruits freelancers through a platform called Neevo and assigns them tasks like labeling images or recording audios hastening their work with Machine Learning powers automation where possible. All the data created or checked by people get compiled into a format that customers can use to train their algorithms.

May Mobility

May Mobility is taking on the self-driving challenge with a form factor that is more predictable than cars. They have developed autonomous shuttles. The company's software has powered shuttle services in Providence, Rhode Island and Columbus, Ohio where you can get a scenic tour of the city as a passenger aboard the self-driving shuttles.

Artificial Intelligence Startup Companies Building Your Smarter Tomorrow

It is not only the big companies who are infusing their products and services with artificial intelligence. Other smaller companies are at work, developing their intelligence technology and services. And well, they are doing a remarkable job. Here is a list of intelligence startup companies you may not know today but which will define your tomorrow in a big way:

Tempus

The company headquartered in Chicago, Illinois majorly focuses on health tech, biotech, and big data. Tempus uses A.I. to gather and analyze massive pools of medical and clinical data at scale. The company uses AI to provide precision medicine that personalizes and optimizes treatments to each individual's specific health needs. It relies on the patient's available data, such as genetic makeup and history to diagnose and treat. Tempus is currently using AI to create breakthroughs in cancer research.

DataRobot

The company is headquartered in Boston, Massachusetts. The company provides data scientists with a platform for building and deploying Machine Learning models. The software helps companies to solve challenges by finding the best predictive model for their data. The company's software is used in Healthcare, manufacturing, insurance, FinTech, and sports analytics.

Narrative Science

The company is headquartered in Chicago, Illinois. Natural science creates Natural Language Generation (NLG) technology that can translate data into stories. By highlighting only, the most relevant and exciting information, businesses can make quicker decisions regardless of staff experience with data or analytics.

Cognitive Scale

The company from Austin, Texas focuses on software and Cloud. Cognitive Scale builds augmented intelligence for the Healthcare, insurance, financial services, and digital commerce industries. Its technology helps businesses increase customer acquisition and engagement while improving processes like billing and claims. Cognitive Scale's products are used by big companies like JP Morgan, NBC, Chase, Macy's.

AlphaSense

AlphaSense concentrates its model in FinTech. It is an AI-powered search engine designed to help investment firms, banks, and Fortune 500 companies find essential information within a large pool of scripts, filings, news, and research. The technology uses Artificial Intelligence to expand keyword searches for the relevant content.

Clarifai

This software company headquarters in New York is an image recognition platform that helps users to organize, curate, filter, and search their popular media. The platform allows images and videos to be tagged, teaching the intelligent technology to learn which objects are to be displayed in a piece of media.

Neurala

Neurala is developing 'The Neurala Brain' a deep learning neural network software that makes devices like cameras, phones, and drones smarter and easier to use. The company's product is currently used on more than a million devices across the globe. Big companies and organizations such as NASA, Motorola, Huawei are also suing this technology.

NuTonomy

The company focuses on developing applications for the automotive and transportation industries. With a mission of providing safe and efficient driverless vehicles, the company is developing software that powers autonomous cars in cities around the world. The company uses AI to combine mapping, perception, motion, planning, control, and decision making into software designed to eliminate drive-error accidents.

Sift Science

Sift Science provides multi fraud management services all in one platform. Sift uses thousands of data points from around the web to train in detecting fraud patterns. The technology helps payment processors, marketplaces, e-commerce stores, and even social networks to detect and prevent fraud. The company's product is used by leading companies such as Twitter, Airbnb, and Zillow.

Zebra medical vision

The company has developed software for radiology and medical imaging which has enhanced the diagnostic abilities of radiologists while maximizing focus on patient care. Zebra works with millions of clinical records and images to create condition detecting algorithms. These algorithms will, in the end, help medical professionals identify high-risk patients' earliest and manage growing workloads with more accurate outcomes.

Sherpa

Sherpa is a virtual personal assistant powered by predictive artificial intelligence. The VPA integrates with the user's entire web of devices, inferring and predicting their needs. Sherpa is continually learning and analyzing more than 100,000 parameters daily to keep the information updated and users organized.

OpenAI

It is a non-profit research company with a mission to create safe artificial general intelligence (AGI). AGI aims at creating machines with general-purpose intelligence similar to human beings. With a focus on long term research and transparency, the company hopes to advance AGI safely and responsibly. The company has received sponsorship from big companies like Microsoft and Amazon.

Tamr

The data management company was born out of an MIT research project to apply machine learning to clean and organize dirty data that is incomplete or inconsistent. Tamr's system automatically identifies sources of data across a company that can be useful together and tags in an employee to instruct the software on how to integrate it. The company has big clients using their software like Toyota, GSK, and GE.

Bossa Nova Robotics

The startup has rolled out big slow-moving robots to 350 stores, including Walmart. The robots help keep the shelves well-stocked. Its systems read price labels for discrepancies and find gaps on shelves do it can alert management about the problems. The robots easily maneuver around the stores and interpret billions of images in a way that is accurate, timely, and reliably.

Pymetrics

This online recruiting platform helps companies find the right employees by looking beyond experiences and sills on the resume. It has over 80 enterprise customers, including LinkedIn, Accenture, MasterCard, and Unilever. The companies use the software to have its top-performing employees complete a set of assessments as part of their evaluation.

Pymetrics gleans critical emotional and cognitive traits for different roles when job seekers apply to work at one of the companies and complete the challenges themselves. They are then paired with jobs that best fit their abilities.

Companies also use the platform for internal career development. It makes the process more efficient with better outcomes and increases diversity in a big way. The company is currently open-sourcing its algorithm auditing tool with the overall aim of preventing its systems from reinforcing gender or bias.

K health

K health has developed a software that enables you to access medical services without having to step into any hospital the company's consumer app draws on a data set of more than 2 billion anonymized medical records sending subtle patterns in data to give users personalized health advice. The company has also partnered with an insurance company, Anthem to let members see how medical practitioners diagnose and treat people with similar symptoms for free. The software also enables you to have a live chat with the doctor but with a fee.

Ways to Invest in Artificial Intelligence

Artificial intelligence is making serious inroads in the investment world. From funds to stocks to money managers, AI is infiltrating the investment landscape in ways no one had to imagine. What makes AI more attractive to potential investors is the fact that AI is touching almost all modern industries, including finance, national security, Healthcare, criminal justice, agriculture, transportation, and many more. Here are some good investment ideas that are highly recommended if you want to be part of the AI action in terms of investments:

Invest in Low-Level AI Funds

There are a wide of AI exchange-traded funds for you to choose from. You can use management fees to help you pick a suitable fund for you. Pick a fund with low fees because this means that more of your money will go to investment rather than to management. Generally, one-year-old iShares Evolved U.S. Technology levies a 0.18% expense ratio and uses a proprietary algorithm to select technology shares. You can get a return of 23% from your investment in a year which is more double the 10.77% category average. For more AI options, iShares offers these low-fee funds: Evolved U.S. Consumer Staples ETF, iShares evolved discretionary spending ETF, iShares Evolved U.S. Financial ETF and iShares Evolved U.S. Media and Entertainment ETF.

Invest in AI Top Stocks

If you prefer investing in stocks and you want to grow your A.I. stock portfolio, then consider investing from these firms:

- Micron Technology, a memory chip manufacturer which is likely to benefit from the demand for increased computing power.
- Nvidia corp.- this is another chip manufacturer with a bright future. The company has been trending lately and is expected to post positive results over the coming years.
- Baidu- this is a massive Chinese company which is investing heavily in AI With about 1.4 billion Chinese citizens, Baidu can capture a tremendous amount of machine learning data to both use and sell. The company promises good returns with the Chinese government deploying AI nationwide.

Invest in Funds that Focus on Robotics

Stock picking is not easy. You may have a challenge picking the losers and winners. If you want to avoid this risk associated with stock picking, then invest in a fund. Investing in funds spreads the risks around. Robotics is a field of robots that are trained to perform as humans. It is a clear example of A.I. in practice. There is two popular robotics fund you can invest in: iShares Robotic and Artificial Intelligence ETF(IRBO) and ROBO Global Robotics and Automated Index ETF (ROBO). ROBO has returned 16.7% to date while IRBO returned19% this year.

Invest in AI Customized Solutions

Big companies are now using AI to understand better and meet their customer needs. Such companies use big data and

AI to understand and analyze shopping, browsing, and demographic data. This allows the company to market directly as well as advertise products and services that reflect the customer's shopping and browsing patterns. You can invest in such AI customized solutions to help improve the running of your business, for example, McDonald has embraced AI solution and technology with ordering kiosks and delivery services.

Invest in AI Startups

Hundreds of new startups focusing on AI research have sprung up recently. While shares of these companies aren't publicly traded yet, you can still invest in them by using startup investment funds like Wefunder, SeedInvest, and 1000 Angels.

Invest in Artificial Intelligence ETFs

More AI-based ETFs are being developed, but there are two ETFs which can help you get diversified exposure at the moment:

- Global X Robotics and Artificial Intelligence ETF (Stock sticker BOTZ)
- Nasdaq Artificial Intelligence and Robotics ETF (stock sticker ROBT)

The Top Artificial Intelligence Stocks for Investment

There are a few companies at the lead working on getting artificial intelligence more involved with your lives. If you prefer to invest in stocks, then these companies are projected to offer handsome returns, in the long run. These companies include:

Google Stock

The company has put incredible resources into many AI projects. Google AI is the company's research branch, and the company has wholly refocused from mobile technology to ai research as its core mission. The company as so far developed innovative AI products such as the Goggle Duplex a sophisticated AI capable of making appointments by telephone with perfectly mimicked human speech. Google is one of the leaders in AI development and is a favorite to investors seeking AI exposure.

Nvidia stock

Nvidia stocks have been at the top of every investor's list of choices. Nvidia, the chipmakers from Santa Clara produces the most potent computer processors in the world. It has leveraged its unique technology in AI research. The company recently had an AI-based breakthrough in brain imaging through a partnership with the Mayo Clinic. Nvidia stock is expected to be on the rise in the coming days.

Microsoft Stock

Microsoft Azure is the most robust AI platform which incorporated a wide range of tools so data can be collected and analyzed better. The new CEO has reinvigorated the company, and the market has responded well to the new changes with investors being rewarded well over the last years.

Salesforce Stock

Salesforce is known for its obsessive focus on seeking growth and opportunities to scale. The customer relationship management software giant regularly acquires hot tech startups to improve its software offerings. In 2019 the company received Bonobo AI, affirm using automated analysis of customer phone calls, texts, and chats to deliver actionable insights. These technological advancements make Salesforce one of the new members on the list of best A.I. companies to invest in.

Amazon.com Stock

The company has a valuation of $1trillion, and it is investing heavily in artificial intelligence. Amazon uses AI to power key capabilities like forecasting product demand, optimizing logistics and warehousing and improving the voice-powered Amazon Alexa virtual assistant. With about 360 AI jobs posted nationwide, the company is at the leading front in the A.I. world. Putting your money in the stock of this company guarantees you handsome returns.

How to Create an AI Start-up

Artificial intelligence is all the new craze right now. If you are wondering how you can build your own AI startup, then here are some brief steps on how you can get started:

Pick a Topic You Are Interested In

You need to select a topic of interest to you to stay motivated and involved in the learning process. Focus on an existing problem and look at the solution. If you are not successfully solving a problem that customers are willing to pay for, then your start up is going nowhere. Before you go far, test whether or not people are willing to pay for what you are planning to build.

Play the Data Gathering/AI building a Game

Once you are assured of ready customers for your product, then build the first generation of your AI. You need to gather as much data as possible touching on your pet project, then curate it to make sure it is useful and then design a model and train it. You need to note that the process of data gathering, curating, and training is the most challenging part of your project. Training a model is very demanding and time-consuming. You will spend a bulk of your problem-solving efforts on gathering and understanding data.

Build your Product

At this point, you should be having a working AI You need to improve your A.I. capabilities to make it user-friendly. This means you have to package your ai into a product which the market can respond to positively. Your product should have a user face with a range of capabilities. Your product must also be able to efficiently solve a problem it was created within a short time and in an affordable way.

Develop a Means for Improving Your AI

Once you launch your new startup, you need to continue gathering more data that was not available for you when you were training your AI You then use this data to improve your A.I. to improve its accuracy. Determine the number of times you will train your AI in a given period.

Finally, in case you are having any trouble building your startup, consider enrolling in the online courses such as:

• **Learn with Google**-this is a project launched by Google to help you understand what A.I. is and how it works. It has a machine learning course for beginners such as GOOGLE's TensorFlow

• **Stanford University Machine Learning**- the course is available on Courser. The course can be offered for free, or you pay a fee if you want a certificate. The course will familiarize you with examples of A.I. driven technologies. You also learn the basics of software engineering

• **Nvidia-Fundamentals of Deep Learning for Computer Vision** -computer vision is a discipline that focuses on creating computers capable of analyzing the visual information as the human brain does. The course covers the necessary technical fundamentals along with practical applications of object classification and object recognition. You will also learn how to build your neural net application

Global Human Machine Industry (HMI)

What is HMI?

Human Machine Interface is a central control system that communicates operator inputs and receives real-time data and feedback from a PLC logic controller. An HMI provides an essential visual of what is going on inside a control system. It records crucial production information, including cycle count times and recipes for different processes. HMI has become the standard interface for operator control on new and upgraded plant equipment and process control system.

HMI and the Global Market

The global HMI market is expected to generate revenues of more than $8billion by the year 2023. The increasing importance of safety, energy, better productivity, and sustainability will lead to the rapid expansion of the global human machine interface market. Moreover, the implementation of supervisory control and data acquisition (SCADA) and Programmable Logic Controllers (PLC) in factory automation systems will transform the market over the years. The global HMI market is driven by growing demand from semiconductor fabrication plants that requires control centers, clean rooms, and assembling plants for effective equipment control.

HMIs facilitate a high degree of customizations and are increasingly finding their application in equipment which is characterized by ruggedness and reliability in the human-machine interface market.

HMI Market – Geography

The global HMI market is divided into North America, Asia Pacific, Europe, and the Rest of the World. Currently, this market is dominated by the region of North America. A rising GDP and recovery from economic slowdown are the main factors driving the North American HMI market. Other factors that make North America dominate the market include the growth in discrete industries, high adoption of advanced manufacturing practices, and increasing demand for advanced software solutions in manufacturing industries.

However, it is expected that Asia Pacific HMI market will witness the highest rate of growth owing to rapidly growing economies like India, South Korea, Indonesia, and China and the growth of automation in these regions.

The companies in the global HMI market are:

ABB Ltd., Advantech Company, Ltd Beijer Electronics, Inc. Honey Wall International, Mitsubishi Electric Corporation, Siemens AG, General Electric Co., Kontron AG, Ltd Rockwell Automation, Yokogawa Electric Corporation.

Estimation of HMI Market

It is estimated that the Global Human Machine Interface Market will expand to US$5,579 million by 2019 as industrial efficiency takes center stage. The market is widely expected to grow at CAGR of 10.4% between 2013 and 2019 with the exception of topping the market with a value of US$5579 million by 2019.

The Global HMI market is also expected to further grow to us$8.96 billion by the year 2026, GROWING at a CAGR of 9.8%. The rising need for efficiency and monitoring in manufacturing plants and the evolution of industrial internet of things (IIOT) and the growing demand for smart automation solution are some of the key factors propelling market growth

Future Technological Advances You Can Look Out For

The technological landscape is witnessing so many advancements nearing maturity. In a decade several of them will be accepted as part of your everyday life. They will influence how you live, work, and entertain yourself. Most will be promoted by business first before being taken by the wider society. While they will make life, easier some will bring fear and uncertainty at the loss of independence and control over your life. Here is a futuristic look at how the world is likely to change in the next decade with technology and innovations.

Flying Taxis

Mobility is expected to change with the introduction of the on-demand flying taxis. An experiment in the area of flying taxis is being orchestrated by ride-sharing provider Uber Technologies as part of its Uber Elevate program. The company plans to deploy Uber Air-the electric aerial vehicles capable of vertical take-off and landing.

Mobile Phone Advancement

A decade from now, mobile phones will have features that you usually see in movies. Such features include the hologram. In the next few years, hologram displays will be commonplace on smartphones.

Your phone will also evolve to become increasingly part of the internet and the Internet of Things (IoT). Future mobile and communication devices will be able to recognize mobile phones uniquely wherever they go shopping or take a pod to the nearest Hyperloop station.

Drones

A future with drones in it is probably closer than you can imagine. There are a million experiments on drone use across the globe. Amazon is currently working on drones which will make deliveries. There are also companies working on drones that will make medicine delivery to remote hilly areas of the world.

Creative Learning in Schools

From the infrastructural point of view, advances in technology will bring to bear many tools for teaching and learning. A good example is a virtual reality which brings alive various concepts, creating an immersive environment in the classroom. With the availability of cloud computing and storage, students will have access to their lessons and worksheets wherever place they are.

Teachers will have to reinvent their roles. They will become catalysts for problem-solving and for connecting students and the industry in the process of learning will become highly interactive.

Medical Services

You will no longer have to go to the hospital for medical services. You can access these essential services right at your home. For example, portable versions of large scanning machines are being built by startups and medical technology companies. These new-generation devices make diagnostics available at home. Moreover, wearable devices that can monitor patients round the clock and upload data to the Cloud securely to the patients' health records are already a reality. They allow doctors to be alerted to a change in a patient's condition and help them determine the severity of the illness.

Chapter 5: 10 Most Influential People And Their Contribution In AI Development

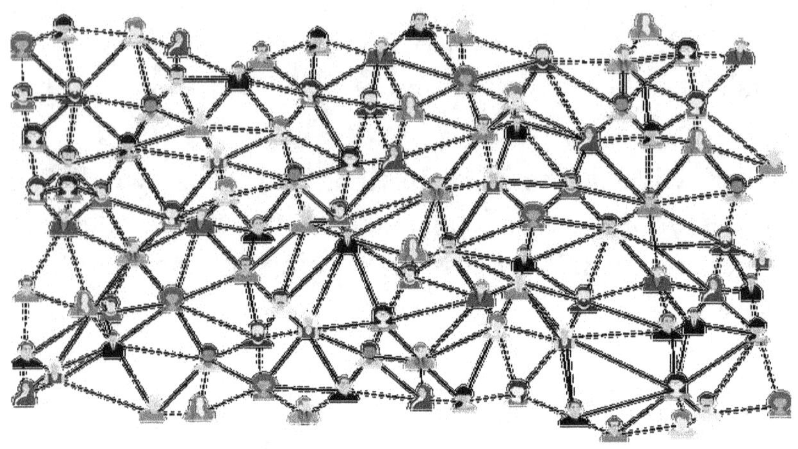

"Technology is unlocking the innate compassion we have for our fellow human beings"
— Bill Gates

As soon as something is working out, then there is no need to call it AI any longer. John McCarthy invented the word Artificial Intelligence in 1995 and ever since the AI Company primarily has undergone through various hard and soft moments. The moments have been accompanied by progression mixed with a lot of promises full of letdowns and disillusion. Coming up of enormous information, quick handling haste, and reasonably improved fever from the American Mafia, AI dignified to create a distraction on a gauge that can exceed the internet itself. As the world is preparing to witness a transformation of the AI companies, every individual associated with the modernization frugality will have to realize how AI will or will not bring impact on their company's daily lives.

Part 1: Company Titans (Superpowers)

The following individuals have helped in shaping and enhancing the AI industry helping it to grow entirely into a thriving industry.

Amit Singhal, Uber

Having been appointed as the new chairperson of the Uber's engineering squad, Amit is going to make a lot of changes to help improve the autonomous automobile breakthrough all over the Uber convoy.

Andrew Ng, Baidu

He is one of the brains behind the creation of Google and now has been appointed as the chief scientist at Baidu," Google of China." Andrew is shaping the AI interaction in the two major economies across the world. There is no harm in him being the co-founder of Coursera, and he has 100K followers on his Twitter account.

Elon Musk, SpaceX, and Tesla

He may not be a researcher of an AI company, but when Elon talks about the industry, a lot of people give him a listening ear. He is the chair of OpenAI. His recent announcement was the launching of a startup, Neuralink, which would create implants that connect the computer interface to the human brain through the use of AI. According to him, this technology would implant minuscule, flexible electrodes to the human brain via a neurosurgical robot.

Jeff Dean, Google

He is the creator of MapReduce, Google Brain, TensorFlow, and Jeff has been considered as the heart of Google's most vital projects and has continued to create waves at the industry who first coined AI-first approach. Together with other Google employees, Dean recently announced the current ongoing AI projects that Google is working on, including Project Diva that assists people to command Google Assistant without using a voice, Live Relay, which is a speech-recognition app meant for the deaf, and project Euphonia, which is aimed to help individual with speech impairment communicate easily.

Ginni Rometty, IBM

IBM ever since Ken Jennings 74 won on Jeopardy in 2004 has been working on Watson. IBM has positioned all its energy and future on Watson. Ginni also wanted to have Watson to tweet for her too. The jury has apprehended Watson but commended Ginni for turning the main focus of the 105-year industry on AI.

Martin Ford

He is the author and creator of Rising of the Robots; he shapes the natural interaction around AI and robots. He is considered to come up with many bestselling books on AI and the future. His recent prediction, according to GE Reports, indicate that AI will destroy a lot of jobs compared to the number of new jobs that will be created. He noted that AI could threaten almost every job that is fundamentally routine in nature.

Ray Kurzweil, Google

The main granddaddy of the association, Ray Kurzweil, has been giving out forecasts which are accurate about Computer Company for many years. He came up with the theory of Singularity, forecasting that in the coming decades we will get to a point where computing power will be so dominant that it will eclipse any reason human attempt process to process the speed of innovation.

Sebastian Thrun, Udacity

He is one of the frontline brains on self-driving cars, Sebastian has also got time to start Udacity. There have been queries on why AI professors from Stanford began to a high online studying industry. This has been because they apprehended the disruption in AI that requires society to re-educate.

Yann LeCun, Facebook

He is the leader of Facebook's group in New York. There is no further explanation about that. Reaching out about its inherent nature means that it will bring a positive impact on AI that will intimate a lot of people. Yann, who has a huge responsibility has come along with massive accountabilities.

Part 2: Rising Stars

The rise of these people will have a significant role. Most of them are not that popular and will contribute positive energy in the coming decade.

Boris Sofman, ANKI

Putting together the power of soft and hardware, ANKI is producing a high class of intellectual customer goods. It was first revealed at the 2010 Apple WWDC. ANKI has various hit goods that incorporate intricate intellect in human sociable robotic puppets.

Bryan Johnson, Kernel

Many people know Bryan from his disbursement startup Braintree, his foray into AI is one of the most daring steps. He is ambitious and thus is the continent vast. They have been working to improve human capacity through direct assimilation using computer technology.

Carol Reiley, Drive.ai

Carol is pushing ahead the autonomous driving uprising. He has been a long-time leader in robotics and is the wife of Andrew Ng. This is the first AI family in the United States to be precise.

Charles Jolley, Ozlo

In case you need a friend then you have not to worry but get Ozlo. The industry has been training Ozlo to be a great and efficient assistant with each interaction and programmer and to create a useful part of your daily lives.

Matt Zeiler, Clarifai

For there to be an interaction between AI and the real world, they need to see the world first. This is where Matt's company Clarifai emanates in. The company helps in operating image and video recognition platform that many AI goods will put in use in years to come.

Nikhil Budama, Remedy Medical

After graduating, Nikhil, together with his team, have been working to enhance healthcare for more accessible and manageable for every person and improving the focus and professionalism of doctors. Remedy Medical has been able to help doctors do administrative tasks and patients to have foster care.

Rand Hindi, Snips

He is believed to be on the race of creating wise technology that disappears in the background and works. Together with all his team of stars at Snips, they are manufacturing artificial assistant that is no-frills and works.

Riva Tez, Permutation Ventures

Riva has been known for being an AI investor and evangelist, bringing problems affecting healthcare and AI into the limelight conversations. After overcoming individual adversities, Riva is now working on changing both AI and the endeavor capital industry.

Shivon Zilis, Bloomberg Beta

Being the curator of the market ecosystem, Shivon brought machine intelligence into the map. You can look for Shivon to influence the direction of the AI Company more so once receiving the support they want to improve in the coming years.

Sinan Ozdemir, Kylie

Sinan was a former data professor at Johns Hopkins University and currently, the founder of Kylie, who is an AI that clones worker characteristics to automate interactions amid the industry and buyers. He is the editor of Principles of Data Science.

Comet Labs Team

In a list of 10, it will be 11, and you can't miss recognizing this team from Comet Labs. They have focused on bringing AI and robotics industries to market, putting together the better of first stage fund boosted lab partnering with big corporations worldwide.

Contributions

Below are some of the impacts on customer experience

The Smart Spread

You have to think about what you in your pocket. Don't think about coins. This is because you are in your particular portal that every piece of humankind's knowledge was formed. Don't marvel at it anymore, but consider your smartphone miraculous. It's a digital helper that can answer questions, plan your life, and connect you to many resources worldwide right away. Don't think it's only your phone, but every technology is surrounding you is growing conveniently. There is a situation where your fridge can alert your milk level is running low; the kettle boils automatically when you wake up, cars dialing emergency services for you, have food by talking to smart speakers. You can have the AI toothbrush that will map your mouth and gives you a report. Nowadays everything has a connection, and the life we are living and the technology being used is that of sci-fi dreams. Just try thinking back how fast and perfect everything has grown comparing it to fifteen years back.

Looking Back at the Past

Can you look back when the internet was readily accessible? Those are the days when you relied on CD-ROMS to get data, or you could go to the library. When phones had no maps, people would be stuck with an A-Z map to navigate. You might not be old enough to think about that, but you can still think about the hard moments of the slow net, tiresome rules on manual devices, and mobiles lacking Wi-Fi connection. You had to wait and be patient. Internet was opt-in and accessing it was to go through stressful processes installing a series of various technologies. Ever since the real-time revolution has been seen happening. This is precisely what it says on the tin, and it's the phenomenon of most things surrounding that is happening in the actual time.

Real-time Revolution

It has made our individual lives simple and has changed the way we communicate with people around the world. With such an improvement, you have expectations of having information or data immediately. This is the first impact of AI on customer experience; there have been high anticipations from technology. You should anticipate your communications with technology to be effortless and anything not fast and giving poor usability sticks out like a sore thumb. When you are still paying attention, real-time revolution creeps in around you and changes your life.

How Are Businesses Coping Up?

Huge Gaps to Fill

Our individual lives are getting more comfortable. This has made it hard for businesses to excite you. You can have a slick,

AI-driven breakthrough at home that you will expect slick, AI-driven advances in your communications with companies. It still doesn't mean enterprises are delivering and giving out their best. All of you will be frustrated like when websites fail to load correctly on mobile when they aren't receiving emails and slowly accessing customer service. It can be wrong to say that change isn't happening. Trades may seem not to be able to serve you as fast as Siri, but starting to be smart due to technology. What are some of the techniques that trademarks are using to keep up?

Chatbots

There are chatbots, and at home, you also have Google assistant, and again when you get online, there are bots. Many businesses are getting chatbots into their websites, thus giving customer operation 24/7 and the whole year. You can look back at the days when you used to get your phone and speak to a business firm. Up to now, you can still find yourself in a FAQ section that takes a lot of time scanning. This may be a time you want to get a refund or update an email address. The addition of the chatbot has improved in ending the slow activities. They give you answers to your queries and are getting smarter in whatever assistance they are supposed to offer.

Chatbots are the best in replying to your FAQ and completing easy account tasks like creating new passwords. When you get the AI-powered chatbots that are being introduced to the market, then you will be advantaged. The new chatbot uses machine studying and natural language dispensation to understand what you mean and your mood sincerely. They can sustain humanized two-way interaction and help you with experience the same as what a real customer service operator offers. Faster and efficiently round the clock interaction has been a significant aspect of the impact of AI on customer familiarity.

E-commerce Personalization

When you look beyond chatbots, websites are significantly getting smarter. Machine studying aiming at technologies are improving and pushing forward wise opportunities for personalized e-commerce understandings. Whenever you see adverts when you load a website, they are not personalized to you and based on other factors or things you like and don't like. Most of the retail websites will applaud goods based on the pages you have visited and viewed and the activities you had on the site. It's usual to see websites greet you by your name nowadays. In case the website doesn't recognize you by your name, it will be aware of your browsing site. There can be an offer basing on your location, or a pop up that will offer you a discount as a new visitor. You won't stop thinking about it, but it's more convenient than where websites were a few years back.

Getting to Know

This is not the smartest stuff to be accessed online. You can see things such as WebVR where you can see, for example how pair of sunglasses look on your face before you decide on buying them. You can see 360 degrees websites that engage you within artistic settings and allow you to communicate with sections like video games. Biometric technology is also rising, and your iPhone cab is unlocked using your face. Trademarks are already attaching this technology for modernized checkouts. This has enhanced the use of fingerprints to finish a PayPal transaction. Soon you will be making transactions that will be verified using facial or voice identification. AI knows people better, thus improving and bringing about the changes seen online.

Oil up Your Onboarding

It's not only browsing that gets more frictionless. Have you looked back about the last time you downloaded an app, digital good, or even a free software trial online? How long does it take to download? Guess you were downloading various pieces of stuff to make it run and did you experience consult instructions? All the tiresome onboarding experiences are things of the past. Trades are well informed that at home, customers are using technologies that don't need brains to activate in spite of authoritative abilities. This has brought about streamlining services offered that can help you run with a good in minutes. As long as you are running, you upgrade by a click or easily handle your account by the use of a secure self-service area. We forgot the days when you had to gather several different information or software to set up a digital good. Everything is coming back to the real-time revolution where people aren't patient at all. People are getting used to better and are expecting better from businesses.

Impact of AI on Consumers

Customer Service Costs

The impact of AI on the customer experience is not perfect. Weatherspoon has allowed customers to order and buys foodstuffs via applications, thus allowing you to sit down and wait patiently for a paid meal. It is easy and efficient; the experienced can be destroyed by a careless waiter that is delivering food to the table. Technology can increase the speed and transactional of tasks, and there is no doubt about it at all. But again, there is an amount of technology that can save your industry from poor customer services. AI can take that risk and break or make your experience. Technology has been considered tremendous, but many businesses can't afford to robotize their workers. You may be in love with slick skills, but customers still want a friendly human feeling.

Rivals or Bros?

You must get to understand why there is love for a human feeling or touch. Machines are wise, but still, they don't follow us as humans do. There is some barrier. There can be jokes shared between you and the robot, but you can share a smile. AI won't be able to access your sarcastic tone or your dry sense of humor. Robotic receptionist won't understand what your raised eyebrows mean, and chatbot won't realize what wow represent. Businesses relying on AI have a risk of alienating their buyers and enhancing horrible robotic experiences. Think of a situation where you are frustrated and want someone to talk to, will you choose a sympathetic human or machine?

Not All Doom and Gloom

There is an open place for AI when it's used subtly to assist service unlike overpower it. Using it well it will impact AI on customer experience, thus being relaxed and productive. AI has pushed us to work all round the clock service. You can visit a company website at any time of the night and ask a question through their bot and have an answer immediately. You will realize that you aren't talking to a human, but still, it won't matter. You can't expect to speak to an operator at midnight you will be happy to have been helped at that time of the night. AI is a big boost for shopping online. North Face is an example of a company that has improved their operations using AI. North Face is assisting online buyers in getting their perfect goods. The use of voice input technology questions buyers and when they want to have their rights. This creates a great experience.

Offline AI

Everything is not about online because AI will assist you to get better services in physical stores. You have to consider geolocation. You have to know where and who allows trades to send you real-time. If you a fan of coffee and it's just around you then there will be an advantage to have an offer enticing you to have your favorite cappuccino. AI has an interest in how humans look. Technology has adjusted adverts depending on time and date. This has brought about great bespoke advertising experience for every individual. Facial recognition has now been captured. You can start seeing offers on your computer screens basing on in-store dwell time.

Fine Balance

This is an example of AI being used as a refreshing addition to streamlining the customer journey. When you are used to this, then the AI influence on customer experience will be impressive and wise.

Impact of AI on Companies

Automation Anxiety

You have seen how AI has an impact on the customer; you can also ask about its effect on the industry testing its implementation. You can be naïve if you claim AI will be leeway and daisies for industries straight away. There has been a series of hindrances to pass through before AI comes up with universal joy in the office. You to put into consideration that the world is getting into automation anxiety, and it causes fear. AI has come up with increasingly sophisticated and affordable developments that humans can replace. Everyone is worried that robots and automated processes will take away jobs from people. This should not be a surprise because AI is taking over even white-collar jobs in marketing nowadays. The tasks robots can perform are enlarging. In the offices, a lot of complicated techniques are being enabled, thus the creation of software that can work automatically executing difficult tasks. With the quick development of AI, then most humans tend to find their jobs at risk.

Morale Dive

It will give a morale challenge to industries that want to introduce AI. Bosses are required to think about effects that could be there when introducing AI on their team. Workers can start panicking and fear the loss of their job. They can begin interacting with others and spread rumors all over the office. It can result in many workers starting to look for other places to work. Implementation of AI could mean traders see workers' morale take a nosedive, and can bring a negative effect on productivity. This is something that most bosses don't consider. C-suite is eventually focused on metrics, and their bonuses are tied to money increasing.

All Change

Coming up with AI in the workplace has brought about a significant shift in the way workplace leadership is conducted. How can human and AI partners together in an integrated office? An increase in the need for workplace leadership training has focused on AI. Bosses are supposed to manage their environs populated by robots and humans effectively; thus, not an easy task to lead such a place. There must be skills that have to be upgraded to bring change. A lot of enhancements are being automated, thus the evolution of jobs. Trades have been busy trying to prepare workplaces for AI.

Trust Matters

The impact of AI on customer experience can bring risk to the trademark. How can AI not be doubted to work the same as a human? When you have an automated function, there will no humanity skills like empathy, emotions, negotiation, and crisis solutions. This can bring about a problem for AI used in the customer context. You must have a requirement that maintains the capability to talk to a human agent and not robots. Some queries may want an agent to think deeply and have empathy for a buyer. You cannot fully trust AI considering this. Company operations may not be ready for AI, and AI may not be prepared too for the integration of company operations. When you handled AI implementations in the wrong way, then you could be in for a negative effect on the company's operations worse than they were. This can disgruntle buyers and reduce the morale of employees.

Flipside

This can be one side of the coin, and risks are coming with AI, and at the same time, some rewards will be gained or incorporated. You will make a big step by saving money by bringing in AI. This is because it has enhanced efficiency and helped industries to manage time, resources, and energy. AI systems can hold up lots of work of workers helping you faster and having informed decisions. AI also comes with some intelligence that corrects human mistakes and has predictive minds i.e., can discuss and tell the number of business intelligence. This can help in enhancing your experience and gives you room to improve your interactions with customers. This shows that AI can assist in making buyers happy and optimize your forthcoming chances. You should also consider the competitive side of it. AI has still not developed that much, and when you start to deploy it, then there are chances of being the future leader. You should be wise enough to make use of technology trends and be an expert in the developing area. Don't consider it to be a negative impact on human workers. From the other side, AI could come up with better levels of jobs and put out the tiresome administrator that comes along the way of production.

A View Of The Future

Physical-World

You can't easily explain how AI will develop, but a lot of experts have seen some sense when tasks start increasing, thus making it easy for a computer to process. In the coming ten years, AI will be able to take over difficult jobs and time-wasting tasks that humans do every day. The impact of AI will rely mostly on cyborgs. Robotics operations are like to increase. In the ordinary world, you can imagine AI in stores that can choose the music to play based on the needs of the customers. AI can also decide wallpaper, artwork that a crowd prefers.

Digital World

Not only will the physical spaces be smart. There are services such as Siri; Google Assistant will continue being active and very soon will be genuine to assist whenever you want whatever the topic. Complex mind blogging prediction, a lot of people will disagree that AI will conquer digital tech out of a 2D screen-imprison plan. Your main programmer interface can become the physical environment surrounding you. Looking at history, you can rely on 2D displays to have a game or communicate with a website i.e., you can see AI and internet advancing; thus, this can be replaced by the environment. You can as well have full-scale school field trips being done in a classroom using the VR-kits, boardrooms where performances show three dimension experience that can be felt. Nobody can tell if the virtual world will end up being better than the real world.

Power of Thinking

When you look beyond, you can see we are getting to a point where AI needs no interface whatsoever. Early this year, you saw Mark Zuckerberg announced that Facebook is working on a direct mind interface that will let you navigate technology just from thought. He is not the only one working on mind interfaces. Elon Musk is creating a brain-machine interface that can connect humans to a computer without any hurdle of input mechanisms. Input has been considered to be a significant blocker, and there have been thoughts that interfaces would break many bottlenecks like has never been seen before. Employees in the offices will one day have to control machines with their minds by sending a thought to an AI system. AI is going to be everywhere, invisible but accessible from our minds.

Chapter 6: AI Applications: The Impact of AI in Finance, Medicine, and Business

"Mining asteroids will ultimately benefit humanity on and off the Earth in a multitude of ways"
— Peter Diamandis

Artificial Intelligence or Augmented Intelligence is significant because it can help tackle primarily troublesome issues in different sectors, for example, manufacturing, retail, medicine, entertainment, transport, and utilities. Artificial intelligence applications can be gathered into these classes:

• **Data**: The capacity to exhibit information about the world. For example, financial management, asset trading forecast, legal assessment, gaming, and autonomous weapons systems.

- **Reason**: The capacity to tackle issues through logical derivation. For example, advertising, marketing, and predictive analysis.

- **Management**: The ability to set and accomplish objectives. For example, stock administration, forecasting demands, predictive support, physical and advanced system enhancement, navigation, booking, logistics.

- **Communication**: The capacity to comprehend written and spoken language. For example, constant interpretation of spoken and composed dialects, continuous translation, AI virtual assistants, voice control.

- **Observation**: The capacity to interpret things about the world by employing sounds, pictures, and other tangible information sources. Examples of this feature are medical diagnosis and treatment, self-driven vehicles, surveillance.

AI in Business

Artificial Intelligence is reasonably the resurrection of computer code. It is a variety of computer code that formulates decisions on its own, that can act even in unforeseen events by the programmers. As opposed to traditional software, AI features a broader latitude of decision-making ability. Those traits build Artificial Intelligence into a precious asset throughout several industries; be it to assist guests and employees develop their approach around a company premises expeditiously, or carrying out a task as complicated as monitoring a turbine to predict its next repairs.

Machine learning is employed usually in systems that capture large amounts of data. As an example, smart energy management systems collect data from sensors mounted to different assets. Machine learning algorithms then contextualize the data troves and hand them over to human decision-makers to higher perceive the use of energy and demands of maintenance.

Many businesses take up computer science (AI) technology to undertake cut back of operational prices, increase efficacy, grow revenue, and improve client experience.

Artificial Intelligence Impact on Business

By deploying proper AI technology, your business could gain the capability to:

1. Save time and cash by automating and optimizing routine processes and tasks.

2. Increase productivity and functional efficiencies

The automation potential offered by AI to today's business activities and functions has gone farther than the assembly lines in history. In many business functions, like promotion and distribution, AI has been ready to speed up processes and supply decision-makers with reliable insight. In advertisements, as an example, the automation of market segmentation and campaign management has enabled a lot of well-structured decision-making and fast action. You get valuable insight into your customers, which may assist you to enhance your interactions with them. Marketing automation is amongst the most useful features in a proper customer relations management application.

3. Make quicker business decisions supported outputs from cognitive feature technologies

There are several complexities to every marketing decision. You have got to observe and understand client desires and needs and align merchandise to those desires and needs. Likewise, having a decent understanding of fixing client behavior is crucial to creating the most effective selling choices, within the short- and long-term. AI modeling and simulation techniques alter reliable insight into your emptor personas.

4. Avoid mistakes and 'human error,' as long as AI systems are created properly

The margin of human error remains more significant than the margin of AI errors. And chiefly, the supply of AI errors is human error. Investment is required to improve error detection for each sort of mistake to mitigate their impacts. As humans and machines both evolve, the chances of the prevalence of new errors will increase (and of traditional errors decreases), that warrants adequate risk-management efforts.

5. Use insight to predict client preferences and provide them more fabulous, personal experience

Organizations are taking advantage of AI because it allows them to present their customers with personalized promotions, that successively will increase engagement, helps to boost client loyalty and improve sales. Another advantage of AI is that it is ready to identify patterns in customers' browsing habits and shopping behavior, therefore enabling corporations to craft extremely targeted offers to individual customers.

6. Mining immense quantity of data to come up with quality leads and grow your client base

Cloud-based AI applications are quite advanced that they will quickly come upon vital information and relevant findings while processing big data. This feature offers businesses insights into antecedently undiscovered data, which provides them with a higher advantage within the marketplace.

7. Increase revenue by taking note of, and increasing sales opportunities

Pioneers value revenue-generating applications over cost-saving ones. And turning into an AI pioneer is the best edge to produce a probably insurmountable competitive advantage.

8. Grow experience by enabling predictive and risk analysis and giving intelligent recommendation and support

Finally, AI is marvelous in the sense that it will predict outcomes supported by data analysis. For example, it detects patterns in client information that show whether or not the merchandise presently on sale is likely to sell, and also the volume at which they are going to do that. It even can predict the timeline for when such merchandise demand can decrease. This information is fundamental in serving a corporation to purchase the proper stock- and within the right amounts.

The most thrust for the adoption of AI in business was a competitive advantage. After that, the motivation came from:
- An executive-led decision
- A particular company, operational or technical downside
- An internal experiment
- Customer demand

- An immediate answer to an issue
- An effect of another project

AI Opportunities for Business

It matters not your reason for considering AI; the potential is there for it to change the modus operandi of your business. All it takes to begin is a broad-minded perspective and a temperament to embrace new opportunities whenever and wherever feasible.

Remember, however, that AI is an emerging technology. As such, it is dynamic and quick-paced and should present some unexpected challenges.

Risks and Limitations of AI in Business

While several business opportunities of AI in the market exist, there also are sure barriers and drawbacks to keep in mind.

One of the significant barriers to implementing AI is data accessibility. Data is regularly archived or is inconsistent, and of poor quality, all of that presents challenges for businesses trying to make value from AI at scale. To beat this, you must have a transparent outline from the point of sourcing the data upon which your AI would depend.

Another key roadblock to AI adoption is the shortage of skills and therefore, the accessibility of technical workers with the expertise and training necessary to effectively deploy and operate AI solutions. Studies suggest seasoned data scientists are sparse as are differently specialized data professionals consummate in machine learning, training smart models, etc.

Cost is another primary consideration of acquiring AI technologies. Businesses that lack in-house skills or are inexperienced with AI typically need to outsource, that is wherein challenges of price and maintenance are come. Thanks to their advanced nature, proper technologies will be high-priced, and you will incur additional costs for repair and in progress maintenance. The processing price for training data models etc. may also be an extra expense.

Software programs require regular upgrading to adapt to the dynamic business setting and just in case of breakdown, introduce a risk of losing code or vital information. Restoring this is typically time-consuming and expensive. However, this risk is not any more significant with AI than with different computer package development. The risks can be managed on the condition that the system is well designed and that you are procuring AI to understand your requirements and options.

Other AI limitations include:
- Implementation times can prove to be extensive, depending on the type of technology you are looking to implement.
- Usability and interoperability with different systems and platforms

If you are deciding whether or not to acquire on AI-driven technology, you must additionally consider:
- Customer privacy
- Potential lack of transparency
- Technological quality

AI Ethical Issues

Intelligent technologies are enhancing our work and lifestyle. As technology becomes a lot more capable, our world becomes a lot of productive. However, despite the development of our lives, technology giants like IBM, Amazon, or Microsoft, as well as the bright minds like Hawking, believe that now could be the time to speak on matters concerning the future of technology and the way it is impacting the planet. These issues are moral problems but also a lens into the long-run risks. So, what square measure these individuals thus disquieted about?

1. The potential of automation technology to present the rise of job losses

The labor business is generally involved with automation. As we have a tendency to evolve and build other ways of automating jobs, we have a tendency to perhaps additionally introduce a lot of opportunities for folks with advanced roles in the future, moving from the manual and physical work, that characterized the pre-industrial movement, to cognitive labor that is appropriate for the new order of society. With job loss, there is also a need to deploy or retrain staff to retain them in other job posts.

2. Fair redistribution of wealth created by machines

The American financial system relies on compensation for contribution to the economy, typically assessed using an hourly wage. The bulk of corporations are still smitten by hourly work that involves products and services. However, by the use of AI, an organization can drastically level the dependency on human beings, and this suggests that revenues can redistribute to fewer folks. Consequently, people who have possession of AI-driven corporations can amass all the wealth.

3. The result of machine interaction on human behavior and machine interaction

While humans are, to some extent, restricted within the attention and kindness that they will use upon another person, artificial bots will channel nearly unlimited resources into building relationships with humans. It is merely the beginning of an age where we will interact more often with machines as if they were humans.

4. The need to deal with algorithmic bias originating from human bias within the information

Though computer science is capable of speed and capability of processing that is way on the horizon of human capacity, it cannot perpetually be trusted to be honest and impartial. Google and its parent company Alphabet are among the leaders when it involves computer science and Artificial Intelligence. As seen in Google's Photos service, where AI is employed to spot people, objects, and scenes, it could, at times, get it wrong. Like once a camera lost the mark on racial sensitivity, or once a software system accustomed to predict future criminals showed bias against folks of African-American descent. We should not forget that AI systems are created by humans who tend to be biassed and judgmental. Once again, if used right, or if employed by people who attempt for social progress, AI will become a catalyst for positive amendment.

5. The security of AI systems (e.g., autonomous weapons) that may probably cause harm

The increasingly powerful technology becomes, the more it is used for criminal reasons as well as good. Security concerns apply not solely to robots made to replace human troopers or autonomous weapons, but also to AI systems that may cause harm if used maliciously. Because these fights will not be fought on the battlefield solely, cybersecurity has become even more necessary. After all, we are coping with a system that is quicker and highly capable than we are by orders of magnitude.

6. The need to mitigate against unplanned consequences, as smart machines are reliably considered to be taught and develop independently

As we tend to build technology, we want to know the mechanism of reward and aversion, that is employed for humans and animals. For robots, these systems are partly superficial. However, they are turning into more elegant systems. In this scenario, might the machine suffer once it is rewarded with negative input? Some genetic algorithms produce multiple instances of a system quickly; however, only the most successful survive, the remainder are deleted.

While you cannot ignore these risks, it is necessary to keep in mind that advances in AI will- for the most part- produce better business and higher quality of lives for everybody. If enforced responsibly, AI has powerful and useful potential.

AI In Finance

Followed by AI's massive success in sectors like retail and manufacturing, it is currently ready to transform banking and financial services. The following are several use-cases that banks will take advantage of without delay.

Finance corporations have faith in computers and data scientists to determine the market's future patterns. Commercialism principally depends on the power to predict the long run accurately, and there is nobody better at the duty than machines which may crunch an enormous quantity of data in an exceedingly short period of time. Machines may learn to look at patterns in data history and predict how these patterns would possibly repeat within the future. Aggressive digital transformation is driven by:

• Relentless competition between traditional banks and agile FinTech's and digital-only banks, which are a magnet for customers with progressive service

• The new target automation, big data, analytics, associated innovation adopted in several sectors of the economy together with the finance needs an agile design to support the digital system

• Customers have gotten increasingly digital and tech-savvy especially those of the new generation, particularly below thirty-five years old need to be online and use all advantages of digital mode of services

In the age of very high-frequency commercialism, financial companies are developing AI solutions to enhance their stock trading performance and increase profit. Digital transformation of banks and commercial organizations is essential to leverage the latest technologies and optimize operative and cost efficiencies.

In-App Banking

Today, nearly all banks have a minimum of one mobile application for carrying out casual banking operations like checking account balance, performing transactions, and ordering new checkbooks, and cards. However, a small number of banks are leveraging AI in their apps to fulfill an outstanding level of customer experience. AI-based chatbots, controlled by natural language processing, can serve banking clients rapidly and proficiently by noting routine inquiries and giving data quickly. Tasks such as opening accounts, transferring money between accounts, paying bills, and process entirely different applications.

Sales Processes

Chatbots are better at responsive elementary queries of shoppers. They will act as a virtual employee. Chatbots have unique algorithms that assist companies to work together with their customers seamlessly with minimal human intervention.

Sales executives have limitations like operating hours and trade information. Intelligent systems like chatbots do not have any such restrictions. It will answer any queries of shoppers no matter the time as long because the data is offered within the system.

The new system stores an enormous quantity of worth and trading data. By exploiting this reservoir of data, the system will create assessments, for instance, it is going to confirm that current market conditions are the same as the conditions from a month ago and predict however share prices are going to be dynamic minutes down the road. Therefore, shareholders may manage to make better decisions supported by the anticipated market costs.

Trading

Today, investment firms are looking to data scientists instead of market experts to see the longer-term direction of stocks. The data scientists produce complicated machine learning algorithms that are capable of finding future patterns within the market by being observant of the trends in data history. These algorithms will consume terabytes of data in seconds and might be trained to spot triggers for anomalies happening within the market. Other than that, individual traders may leverage AI to form choices (for them) like when to buy, hold, or sell a stock.

Compliance Control

Every financial establishment faces a high level of scrutiny. An enormous volume of information is being created each minute by banks. It takes months to spot malpractices like money laundering, market manipulation, foreign regulatory compliance, concealment, etc.

Risk Analysis

Artificial Intelligence techniques are applied to risk analysis due to its ability to handle uncertainty, incomplete and inexact specifications, vagueness, and qualitative data.

Artificial intelligence and risk management align once there is a necessity in handling and evaluating unstructured data. It is calculable that risk managers of financial establishments focus on analytics and stopping losses in a very proactive manner supported by AI findings, instead of creating time in managing the risks inherent within the operational processes.

AI solutions are ready to fuel financial organizations with trusted and timely information for building competency around their client intelligence and the imminent implementation of their methods.

Predictive Analytics

As the name suggests predictive analytics, "predicts" a customer's future status. The AI rule predicts what is going to be a customer's status if they will still pay and invest cash just like the manner that they are doing. The machine algorithm can also act as a private financial consultant to a client by providing them a recommendation on how they can improve their status.

Data Enrichment

Transaction data is often robust for someone- such as a merchant- who spends and receives cash multiple times every day. Transaction data transforms the difficult-to-understand transaction information into easy-to-understand by sorting transactions into categories. It helps customers monitor things like credit rating, budgeting, spending habits, analyzing, and predicting the earnings and spendings of the future.

Fraud Identification

Both the amount of worldwide transactions and amount in an exceedingly single global transaction is multiplying like a virus; so is the threat of online fraud. The standard online fraud detective algorithm took merely a couple of data points into consideration, whereas AI-based algorithms consider far more data points to spot fraud. Fraud detection is a notable use of AI in financial services. For instance, MasterCard uses Decision Intelligence innovation to examine different data focuses on recognizing fake transactions, improving continuous endorsement precision, and reduce unnecessary transaction delays. Machine Learning algorithms helped the giant payment and technology, MasterCard, to cut back fraudulent activities by fifty percent.

Banks already likely have all of the transaction information tagged because of their storage of bank records from years past. Fraud consultants at the client bank engaged in the machine learning model have got to label the fraudulent transactions and those that are not when the system is being trained. The software system bit by bit improves at discerning between fraud and bonafide banking operations because it is exposed to a lot of tagged transactions.

Smart Loans

Banks offer loans to their customers, supported by a credit-scoring system. It takes into consideration their banking history, income, tax payments, and more. But, for the customers who have all their money information well-recorded get the upper hand. However, a majority of loan seekers who are underbanked do not have their financial information in bank records.

The new AI-based credit classification system can collect wealth knowledge from the smartphones of underbanked customers to spot their trustworthiness. These different data points can offer new ways for these customers to access credit from the credit banks.

Wealth Management

Play store and app store offer a variety of wealth management applications that facilitate customers to manage their wealth. These apps leverage a customer's checking account details. Banks have the foresight to steal their customers by introducing an in-app personalized wealth management system. These AI-powered consultants ceaselessly learn from our money activities and supply the most straightforward recommendation to customers, the same as what a relationship manager would do.

A digital revolution is returning within the banking and finance sector with AI. The aim of introducing AI in banking and finance is to deliver pleasant client experiences. It helps customers draw their attention away from understanding numerous banking slang and processes, and instead focus on those things that matter most to them. Specialists predict that the future of AI-based banking mobile app development services is brighter than most people assume.

Financial services organizations use AI-based natural language processing instruments to examine brand opinion from web-based interests and give significant recommendations.

AI In Medicine

The healthcare sector has been amongst the highest adopters of AI technology. It boils all the way down to the ability of AI to compute numbers fast and learn from historical information, that is essential within the business.

AI can give information-driven clinical decision support (CDS) to doctors and emergency clinic staff preparing for expanded revenue potential. A subset of AI, ML, is intended to distinguish patterns, utilize calculations and information to give automated bits of data to healthcare providers.

For example, Cambio Health Care developed a clinical decision support system for stroke detection that may offer the medical practitioner a warning when there is a patient in danger of getting a stroke. Another such example is Koala Life that may be a company that has developed an AI-powered device that is capable of noticing internal organ diseases, like cardiac arrests.

Artificial Intelligence and ML innovation has been especially helpful in the healthcare industry since it produces tremendous amounts of data to prepare with and performs calculations to spot designs quicker than human experts. Take, for example, Medecision built an algorithm that recognizes eight factors in diabetes patients to decide whether hospitalization is required.

Clinical Application

The best uses of an AI are augmented intelligence that permits doctors and nurses to perform their best by providing them with timely, data-driven recommendations that they will agree with, reject, or modify supported their personal experience and their judgment of context at that moment in time.

Keeping Well

One of AI's most prominent potential advantages is to assist individuals to stay healthy so that they do not have to visit a doctor, or not as often. The use of AI and also the Internet of Medical Things (IoMT) in public health applications is already serving to individuals.

Technology applications and applications encourage healthier behavior in people and facilities with the proactive management of a healthy modus vivendi. It puts individuals at the top of their health and well-being.

Also, AI will increase the power for healthcare professionals to better perceive the regular patterns and wishes of the individuals they take care of, and in addition to that understanding, they are then ready to offer better feedback, steerage, and support for staying healthy.

Detection

AI is already being applied to observe diseases, like cancer, additionally accurately and in their early stages. Consistent with the American Cancer Society, a high proportion of mammograms yield false results, meaning one in a pair of healthy ladies is told that they have cancer. AI is sanctioning review and interpretation of mammograms thirty times quicker with ninety-nine percent accuracy, reducing the requirement for excessive biopsies.

Diagnosis

Watson for health, by IBM, helps healthcare organizations apply cognitive technology feature to unlock vast amounts of health information and power diagnosis. Watson will review and store much more medical data- each medical journal, symptom, and case study of treatment and response around the world- exponentially quicker than any human.

DeepMind Health, by Google, is functioning in partnership with doctors, researchers, and patients to unravel real-world health issues. The technology combines machine learning and systems neurobiology to create powerful general learning algorithms into neural networks that mimic the human brain.

Treatment

Beyond scanning health records to assist physicians in establishing inveterately sick people who may also be in danger of succumbing to an adverse episode, AI will facilitate clinicians take a new comprehensive approach for disease management, better coordinate care plans and assist patients in raising, managing, and adjusting to their long-term treatment programs.

Robots have been employed in medicine for over thirty years. They vary from easy laboratory robots to extremely advanced surgical robots that may either aid a person's medico or execute operations by themselves. Not only for surgery, but they are employed in hospitals and labs for repetitive tasks, in rehabilitation, physiatrics, and support of these with long-term conditions.

Functional Applications

Hospital operations are complicated, extraordinarily different, and deeply interconnected systems — this can be why it is operationally challenging to deliver high utilization of assets, low wait times for patients, and an oversized variety of available slots for patients seeking a rendezvous, all at the same time. The daily operations need many choices being created by the hospital administration staff levels on a regular basis. Sadly, seldom do days go precisely as arranged, and as a result, the battlefront is forced to suppose a static dashboard, half-formed predictions, or gut-feel to create these choices.

Decision-Making

Improving public healthcare needs the alignment of massive health data with appropriate and timely decisions, and prognostic analytics will support clinical decision-making and actions in addition to prioritizing administrative tasks.

Using pattern recognition to spot patients in danger of developing a condition- or seeing it deteriorate thanks to lifestyle, environmental, genomic, or different factors- is another space where AI is setting out to take hold in medicine. Many data analytics systems concentrate on "descriptive analytics" that answer questions like "what happened." AI permits the analytics to maneuver more toward "prescriptive analytics" that makes recommendations regarding "what ought to happen." The intent is to leverage the information and continuous machine learning to assist the battlefront systematically create the most privy choices across all sorts of circumstances.

Patient care and operations are two totally different, everyday samples of ways in which AI will disrupt the healthcare business. There is an infinite number of additional examples, and nevertheless, hopefully, these demonstrate the very fact that AI harnesses unimaginable amounts of data to boost the expertise in both large and smaller ways.

Research

The path from the laboratory to the patient may be a long and expensive one. Consistent with a Golden State medical specialty analysis Association, it takes about twelve years for a drug to travel from the research laboratory to the patient. A measly five in five thousand of the medication that begins presymptomatic testing ever reach human experimentation and only 1 of those five is ever approved for human usage.

What is more, on average, it will cost an organization US $359 million to develop a brand new drug from the laboratory to the patient.

Drug research is among the newer applications for AI in healthcare. By leading the latest advances in AI to shape the drug discovery and drug repurposing processes, there is the potential to shred both the cost and time to market considerably.

Training

AI permits those in training to travel through realistic simulations that straightforward computer-driven algorithms cannot. The arrival of natural language and therefore the ability of AI computers to draw instantly on an oversized database of eventualities, means that the response to queries, decisions, or recommendations from a trainee will pose a more significant challenge that a person. Therefore, the training program will learn from previous responses from the student, which means that the problems are often regularly adjusted to satisfy their learning desires. And training is often done anywhere; with the ability of AI embedded in a smartphone, fast catch up sessions, after a problematic case during a clinic or in motion, are possible.

Optimization models that learn and adapt will facilitate the improvement of the productive capability of the various valuable assets in a health system. These assets include operation theaters, patients beds, and imaging instruments, to name a few. AI will dictate the fate of medicine as a result of its potential to enhance patient care, bring on higher outcomes, and enhance the expertise, whereas utilizing better the existing resources. The resources at hand help doctors, schedule groups, facility managers, and ultimately even insurance corporations. As additional AI-based solutions are adopted and placed into play, expect an entirely new world of opportunities to open up, which will remodel healthcare for the better.

Conclusion

Thank you for making it through to the end of *ARTIFICIAL INTELLIGENCE AND MACHINE LEARNING: AI Superpowers and Human + Machine A Visionary Revolution in Finance, Medicine and Business. Find Out 10 Most Influent People of the Era With A Modern Approach.* I hope that it was informational and was able to provide you with the basic knowledge you need to understand the concepts of Artificial Intelligence and machine learning. By finishing this book, you will be able to possess the mastery that you seek in understanding the role of artificial intelligence in influencing human behaviors in different facets of life.

We have gone through the definition, goals, and types of artificial intelligence and machine learning. This book has offered easy-to-use but very powerful and effective definition of concepts that are crucial in understanding artificial intelligence. It provides a great overview of how the world is gradually adapting to the technological changes with the aid of artificial intelligence and machine learning. You are now familiar with the relationship between AI and machine learning, as well as the possible differences. You have also learned that almost every aspect of our careers are gradually embracing AI for efficiency.

You are now aware of the key concepts of AI and machine learning, and how they work. The next thing you would want to do is to decide to invest in the field due to the vast opportunities available. With the knowledge of superpowers in AI and machine learning development, you can further learn the more opportunities that exist in such technology.

Finally, if you found this book useful in any way, a review on Amazon is always appreciated!

Artificial Intelligence Business

Artificial Intelligence a modern approach. Data Science, Data analytics and ML applied in Healthcare, Marketing for Real World. Successful business tools in Practice

INTRODUCTION

Congratulations for purchasing your copy of "Artificial Intelligence Business - Data Science, Data analytics and ML applied in Healthcare, Marketing for Real World. Successful business tools in Practice a Modern Approach Handbook". I'm excited that you have chosen to immerse yourself in the interesting world of Artificial Intelligence to help you in your business expansion. You will soon make discovery that there are various Artificial Intelligence applications that are in use in the modern world these apps are designed to help in making our everyday simpler and more productive.

Artificial Intelligence refers to the capability of digital computers or robots that are computer-controlled to carry out tasks normally linked with beings with high intelligent levels. AI is commonly used in the development of systems that are gifted with intellectual processes that characterize human beings i.e. the ability to discover meaning, learn, or generalize from experiences from the past. Demonstrations have shown that AI programmed computers or machines are well capable of handling very complex tasks.

Artificial Intelligence can be referred to as a concept in which a product, a robot, or a computer is made to think in a similar or a better manner than human beings. Artificial Intelligence (AI) can be defined as the study of how the human brain thinks, learns, makes decisions and works, in the process of problem solving. The main goal of AI is in the improvement of computer functions directly related to the human knowledge i.e. problem-solving, learning and reasoning

The main goal of Artificial Intelligence even as it continues being incorporated in business is to have the ability to adapt to and learn from broader array of challenges. This will eventually eliminate the risk associated with limitation in regards to AI problem solving capabilities.

At the same time, it's important to remember that as much as there is the positive aspect regarding how Artificial Intelligence impacts business, AI also has some risks associated with it. These risks can be very detrimental in your business though many technocrats argue that the benefits outweigh the disadvantages.

Artificial Intelligence isn't targeted to act as a total replacement of human beings. AI only serves in augmenting our abilities as well as enabling us to perform our tasks in a better and more efficient manner. Machines that are fitted with AI applications learn in a different way as compared to human beings; this means that they analyse things in a different way. They are better placed to capture patterns and relationships that escape us.

Artificial intelligence does an incredible job in business as far as delivering accurate results is concerned. This is simply because machines aren't affected by external factors as they carry on their duties as is the normal occurrence in human beings.

The most fascinating aspect is the use of AI systems in the medical sector; techniques from object recognition, image classification and deep learning are now being employed to spot cancer with similar accuracy as radiologists (through MRIs). Generally, there are several benefits that are directly linked to Artificial Intelligence as well a number of risks associated with the same. The main impact in the business world is the aspect of time saving and simplifying complex tasks.

There are several books on Artificial Intelligence a Modern Approach, thanks once again for picking this one! Every effort has been made to ensure that it delivers as much information useful for your business as possible.

OVERVIEW

Artificial intelligence (AI), is a concept whose invention goes back to 1956. Thanks to improved computing storage and power, advanced algorithms, and amplified data volumes,
AI has become increasingly popular in the present day.
AI is best described as the human intellect processes simulation by machines. These processes are inclusive of learning, reasoning and self-correction. Specific Artificial Intelligence applications include machine vision, expert systems and speech recognition
Artificial Intelligence enables machines to take to new environments, learn from past experiences, and to mimic human beings in performing their tasks. Most AI models that are in place today i.e. driverless cars, greatly rely on language processing and deep learning. By making use of these applications, computers have been empowered to perform particular tasks by recognizing data patterns and processing huge data volumes.
Artificial Intelligence automates learning that is repetitive and discovery via data. AI is very dissimilar from robotic automation that is hardware driven. Instead of opting for the automation of manual tasks, Artificial Intelligence performs huge volumes of computerized tasks reliably and frequently without getting tired. This automation type requires human inquiry in setting up systems and asking the appropriate questions.
Artificial intelligence is very important in business in that it enhances intelligence to the existing products. In other words, products that you are already using will be upgraded with AI skill. Smart machines, bots, conversational platforms and automation can be joined with huge data volumes; this combination improves several technologies in the business world i.e. from investment analysis to security intelligence.

Artificial Intelligence is a program that makes adaptations through learning algorithms progressively; this allows data to carry out the programming. AL looks for regularities and structure in data which enables skill acquisition for the algorithm. This algorithm then becomes a predictor and a classifier.

Artificial Intelligence applications come in handy in analysing deeper and larger data volumes; this is accomplished via neural networks which contain several hidden layers.

Artificial Intelligence is presently being used in various sectors in business including:

Health care

AI programs have the ability to provide x-ray readings and medicine prescriptions that are personalized. Personal health care helpers can actually step in as life coaches; you are promptly reminded to eat healthier, take your medicine or exercise.

Retail Sector

Al offers virtual shopping abilities that provide recommendations that are personalized and engage the consumers in discussions concerning available purchase options.

Other business sectors in which Artificial Intelligence plays important roles include; manufacturing and banking sectors. Researchers have found out that Artificial Intelligence is one very vital component in the running of business in the modern world. Several organizations have incorporated AI in their systems since AI has several positive impacts in business. Certain concerns have been raised as far as Artificial Intelligence is concerned. However, it has been impossible to overlook the benefits of AI in businesses as a whole. Prior to installing AI systems in your business, take time to find out whether the pros and cons of AI will work for or against you.

CHAPTER 1:
WHAT IS ARTIFICIAL INTELLIGENCE IN BUSINESS?

Prior to making any observations with regards to the manner in which technologies related to Artificial Intelligence are impacting the business fraternity, it's of great importance to have a good understanding of the Artificial Intelligence terms. AI is a general and broad term which denotes to any form of software in computers which performs humanlike activities. These activities may involve easy to complicated roles including planning to more complex ones like problem solving.

Artificial Intelligence is defined as the process in which machines replicate the processes of human intelligence. Some of the processes associated with human intelligence include self-correction, reasoning (making use of rules in reaching conclusions that are definite} and learning (process of acquiring information as well as rules that govern making use of these information). The Al has precise applications including; expert systems, speech recognition and machine vision.

AL is considered to be either strong Artificial Intelligence or weak Artificial Intelligence. Weak AI can also be referred to as the narrow AI; this is a system aimed and designed for precise tasks i.e. Apple's Siri. Strong AI is also known as general intelligence, a system that comprises of indiscriminate human intellectual abilities. When a task that isn't familiar is presented, the strong AI is tasked with finding a solution; this doesn't require intervention from human beings.

Since staffing, software and hardware costs for Artificial Intelligence can be very costly, many dealers are adding AI modules in their typical offerings. AI is also made accessible in the service platforms; this means that companies and individuals are permitted to try out with Artificial Intelligence for certain multiple platforms and purposes prior to making any commitments.

Some of the popular Artificial Intelligence cloud are as follows:

Google AI

IBM Watson Assistant

Amazon AI

While the tools in Artificial Intelligence present a variety of functioning capabilities for many businesses, their use has raised ethical concerns. This is because these machines are programmed by human beings, which leads to the belief that there is bound to be a bias which needs close monitoring.

Some technocrats have come to the conclusion that Artificial Intelligence is a word that is linked closely to common culture; this has caused fears that are unrealistic in human beings concerning AI and the questionable expectations concerning how life and the work place will experience change. Machines will only react and act like human beings only if they are well equipped with adequate information in relation to the present world.

Artificial Intelligence is a component in computer science in which the development machines that are gifted react and perform tasks like humans is emphasized. Some of the tasks intended to be performed by computers that have been installed with A.I are:

A. problem solving

B. Planning

C. Learning

D. Speech recognition

E. Ability to move and manipulate objects

F. Perception

Artificial Intelligence can be termed as a computer science branch whose aim is creating intelligent machines. This particular branch has become very vital in the technology industry.

How is Artificial Intelligence Applied in Business?

Artificial intelligence has been known to be of great impact in the business sector for many years now; it's been in practice for decades. Nevertheless, due to the possibility of accessing huge amounts of data as well as increased processing speeds, Artificial Intelligence has also started taking root in our lives on a daily basis.

From image or voice recognition and language generation to driverless cars, machine learning and predictive analytics, AI systems are applied in various areas. The AI technologies are crucial in yielding innovation, reshaping the manner in which companies carry on with their operations and in the provision of new opportunities in business.

Artificial Intelligence is speedily becoming a tool that is being used competitively in business. It's very clear that companies are beyond negotiating about the advantages and disadvantages of Artificial Intelligence. From data analytics to customer service to arriving at predictive recommendations, AI is viewed by leaders in the business world as a very necessary tool.

Artificial Intelligence has been ranked among the top technologies that a company needs to utilize; this can be done by discovering how to make use of the AI technology to their advantage. There are several devices that we use in our everyday lives that are enabled by the Artificial Intelligence applications i.e. smart assistant devices or apps like Apple's Siri or Amazon's Alexa.

Evaluating How Business Leaders Make Use of Artificial Intelligence in Competing Against Each Other

A good idea is making use of the internet and so the relevant research concerning how other companies make use of AI technology for their advantage.

You can also read through your competitors' social media platforms and websites i.e. Facebook and LinkedIn. You can also browse their blogs, news coverage, and press releases. You can also look manually for annual reports or newsletters that may not be posted online.

You can cast the net wider and do a search i.e. how hotels using Artificial Intelligence, or even 'how companies are using Artificial Intelligence.

Researching on the manner in which additional components of your chain of supply and support make use of Artificial Intelligence. Don't forget to research about other avenues that are non-digital. If you're attending an event that is related to your industry, look out for Artificial Intelligence sessions. Go ahead and speak to people sitting or standing next to you. You could also read journals which will enable you to find out as much as you can about your competitors. Be sure to get information on how they make use of AI to their advantage. Always note down any useful information that can help you in the implementation of Artificial Intelligence in your business. There are several types of Artificial Intelligence applications used in the business world but the most commonly applied one is the Machine Learning software. This particular application is basically tasked with processing huge data volumes, at a fast pace. This could also mean the reduction in the need for manual labour or even result in the over reliance of AI in performing various tasks in the business environment.

How to decide whether Artificial Intelligence will work for you in your Business

Several factors can act as your guideline when it comes to making the decision as to whether to make use of AI in your business. You need to have prior idea based on your findings concerning what AI has done for other companies.

The companies you base your research on need to be in similar industry, and if possible, similar size.

Understanding that various tasks require certain data types to work will enable you to make better decisions on the AI system to acquire for your company. Make sure that you fully grasp the limits and requirements of the intended tasks.

Be sure to consider the types of process that can be carried out by Artificial Intelligence. This gives you an understanding of what AI will be utilized for, ensuring that the choice you make will yield positive results for your business.

Is Artificial Intelligence a Blessing or a Threat in Business?

Technology is a very vital component in the growth and development of humans. AI is among these very vital technologies which are gaining hype and momentum. Technocrats have argued that AI can be a disaster and for some, a blessing.

Some of the advantages of Artificial Intelligence in Business include:

Using Artificial Intelligence decreases chances of making errors

This is mainly due to the fact that decisions made by machines are greatly influenced by data that had been recorded previously hence the chances of any errors occurring is greatly reduced. This can be termed as a great achievement since problems solving can be handled without leaving room for errors, even for the complex ones.

Business establishments that are advanced prefer using digital assistants in the daily interactions with their viewers;

this plays a key role in time saving. This also helps businesses in fulfilling the demands from their users without any delays. These systems have been programmed to deliver the best assistance possible to users.

Artificial Intelligence facilitates appropriate decision making

Machines have absolutely no emotions which makes it very easy for them to make decisions that are more efficient, within shorter time frames. This can be best outlined in the health sector. Integrating Artificial Intelligence in the healthcare facilities has greatly improved efficiency in the administration of treatments. This is through the minimization of the occurrences of inappropriate diagnosis.

Artificial Intelligence can be used in risky instances

There are some instances where safety for human beings becomes vulnerable. i.e. survival for human beings on ocean floors is very difficult. Machines that have been fixed with algorithms that are predefined are used in studying these ocean floors. This is a major limitation that Artificial Intelligence has assisted to overcome.

Machines have the capability of working continuously

Unlike human beings, machines don't suffer from fatigue, even after working for many hours. This is a huge advantage over humans; they require time to rest for them to work efficiently. However, the efficacy of machines isn't determined by external factors. This means that these external factors don't reduce the working efficiency of machines.

Some of the disadvantages of Artificial Intelligence in Business include:

Implementing Artificial Intelligence is costly

When totalling the repair, maintenance and installation costs of Artificial Intelligence, the proposition is clearly very expensive. This isn't viable for industries and businesses that don't have sufficient funds; it proves very difficult for the implementation of AI technology into the businesses' strategies or processes.

Using the Artificial Intelligence has greatly increased the reliance on machines

Human beings have become increased dependent on machines. In the coming times, it might get to a point where it will be impossible for human beings to perform their duties without the aid of machines. The dependency of human beings on machines is bound to increase in future. This will result in the cognitive capabilities of human beings decreasing with time.

Artificial Intelligence has resulted in the displacement of people from manual jobs

This is the main worry for technocrats. There is the possibility that Artificial Intelligence will end up displacing several low skilled workers. Machines have the ability to work non-stop which has resulted in companies preferring to use machines instead of using manual labour. The world is edging towards automation which will mainly result in the majority of jobs being carried out by machines i.e. in the scenario where there is the inception of the driverless cars, uncountable drivers have been pushed into unemployment.

Machines Fitted with Artificial Intelligence perform restricted tasks

These machines have been programmed to perform certain duties with regards to what they have been programmed and trained to do. Relying on these machines in instances that may require the adaptation to unfamiliar environments can be frustrating. Machines don't possess the ability to think in a creative manner since their reasoning is restricted around the specific areas they have been programmed to do.

In conclusion, rather than replacing human ingenuity and intelligence, Artificial Intelligence is generally viewed as a subsidiary tool.

Irrespective of the fact that Artificial Intelligence is finding it hard to complete obvious tasks presently, it's skilled in analysing and processing volumes of data quicker than the human brain. This is a good thing in relation to the business world in the sense that AI accelerates the rate at which tasks are carried out, saving time that can greatly be reflected in increased sales.

When it comes to making the decision about whether to make use of Artificial Intelligence applications in your business or not, always evaluate whether the AL will work for the good of the business or vice versa. Always make considerations about the cost implications as well as other factors like if the systems will have any impact in profits maximization. Will Artificial Intelligence help you in making informed decisions? Will the AI lead to rendering people who have worked for you for decades jobless? What impact will the same have in their livelihoods? These questions will enable you to make informed choices as you take the next step in installing AI systems in your business.

Chapter 2:
A MODERN APPROACH ON ARTIFICIAL INTELLIGENCE IN 21ST CENTURY

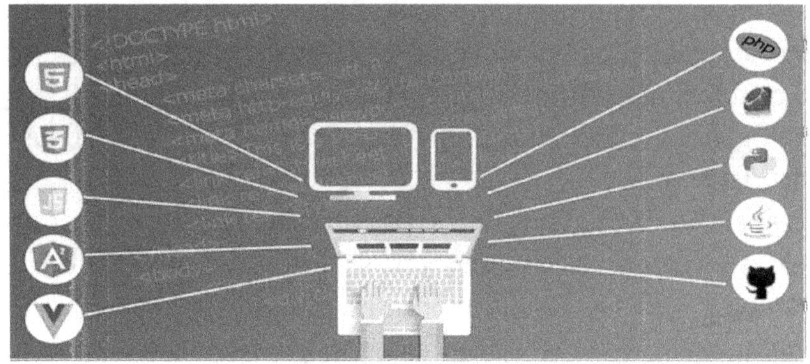

Learning is very fundamental, and particularly for the 21st century human being. Just when you think you have acquired enough, something else comes about, which poses a challenge for us to be on the move always. Essentially, when you discover new trends, your learning mode degrees change. Your level could move from surface to deep, depending on your interest and curiosity. Loosely speaking, as we seek to improve and get deeper into learning, the machines have also not been left behind. They are out to mimic human intelligence and they have proved to do even better than humans in some areas.

It has been many years of rule-based structures. Fortunately, with continued learning, techniques like deep learning have been discovered. Deep learning is one of the approaches to ML that have evolved quite swiftly. However, although it heavily relies on data and looks brilliant, the truth is that there are still numerous concerns that need to be addressed. We need to ask ourselves questions like 'how closely can technology imitate human intelligence?'

Exactly How Artificial Intelligence Resurgence in 21st Century Feels Like

Today's AI systems are not fully autonomous as yet. Even so, they are amazing as they can enhance the competence of human operators. The collaboration between artificial intelligence systems and the techies has brought about better experiences for both the customers and employees in the business setting. From the experts' desk, AI and its ML subset are on the verge of penetrating the military arena. There are proven military technologies that are falling into place, and they offer nothing short of an epic future for the industry.

AI as a modern approach to military technology

Artificial intelligence is one of the primary components predicted to impact war-fighting technologies. You may have heard of the Third Offset Strategy, a technique that the U.S forces use to detect asymmetries between them and their potential opponents. Being in the 21st century, military innovation is under watch for what it will do to enhance defense advantages for U.S forces. The key to the success of the Third Offset lies within artificial intelligence and autonomy. Employing these two aspects in the battle network will perfectly fit into the framework. It only requires one to appreciate the fact that several techniques have to be integrated.

Thankfully, the security and intelligence systems are already in the process of studying such systems as robotics and missile defense for deployment when it's the right time. U.S forces are not the only ones delving into military AI. Their near-peer opponents and competitors are also out to find a way of gaining a corresponding war-fighting advantage against them. Now, such competitive engagements will most likely result in dependence of inscrutable algorithms, which depend on cloud-level computing to determine the fate of the army. As we ponder on this, the most important thing is to prioritize professional military education to wield the said technologies. This can be achieved through democratizing experimentation and augmenting collaboration.

Artificial Intelligence: the Insights you Need
With a lot of improvement being experienced in the AI space, expert insights are always burgeoning. This is an interesting field where remarkable innovations can be discovered by any devoted researcher. As you read, keep your mind open to allow room for independent thinking as there is still more to be done before we can boast about attaining absolute success. Meanwhile, we will thrash out some of the insights that you need for your knowledge.

Understanding the position of data lineage: this is a crucial aspect that maps out the data's course. As you study AI and its techniques in profundity, remember that lineage has significant implications on each of them. These include neural networks, ML, deep learning, and natural language processing. Frankly, it's impractical to have accurate data-feeding artificial intelligence without data lineage.

Artificial intelligence in manufacturing: manufacturers are getting immersed in data daily, especially with the proliferation of networks and sensors in the working environment. We must appreciate that even the apparently negligible improvements have major implications. Artificial intelligence naturally finds affinity with manufacturing because the industry is heavily data-reliant. The truth is that we are on the threshold of a momentum shift.

AI use in shunning fraud in insurance application: submitting your insurance policy applications online sounds like such a deal, right? It is not only convenient but also quick to research on the rates and settle the entire process in a matter of several computer clicks. Sadly, fraudsters are into it too. They perform some malicious tasks like opening policies for imaginary beneficiaries, opening and revoking policies to make additional benefits and quota as well as applicants fabricating information to lessen their premiums. Now that they possess loads of data, insurers can employ AI to verify the authenticity of every applicant.

AI in aiding proper prescriptions and combating illicit drug addiction: it is heart breaking to learn that there are thousands of people who die from prescription overdose every year. According to a report by the HHS, deaths resulting from drug overdose are manifested across the whole demographic spectrum. Opioid overdose has been notorious, and the upward trajectory is quite worrying. The techies can employ data-driven techniques to help save lives. Integration of pattern recognition, ML, and anomaly detection will help analysts to make remarkable advancements like;
Detecting early addiction signs in patients.
Predicting and impeding drug trafficking by quickly detecting suspicious prescription and dispensation patterns.
Ensuring clued-up prescription choices are made. This can be achieved through coordinating treatment whereby patient and drug details are passed to the doctors and prescribers directly.

Artificial intelligence insights and applications are not limited to the above batons. There are other fields of influence like incorporation of chatbots in analytic applications, employing AI in banking, marketing, drug development, and many more. The good news is that even with evident gaps in the AI strategies and capabilities, the industry leaders point out a potential outburst of its adoption. Government leaders and technologists should step up and learn from experiences by their peers. This is the only way through which we will all set out on a successful path.

CHAPTER 3:
OVERVIEW OF DIFFERENT PROGRAMMING LANGUAGES: MODERN AND OLD APPROACH OF AI TECHNIQUES

Computer programming languages are formal languages comprising a compilation of instructions that generate a range of outputs. They are primarily used in implementation of algorithms. These languages provide the most ideal platform for programmers to develop applications, games, software, etc. Some of them can be integrated while others can be considered to be independent. We cannot deny that this is an indispensable point to start from even as we scale to greater heights of machine involvement in performing human tasks. Programming languages are divided into 3 main kinds defined below.

1. **Machine language:** this is a set of binary digits that a computer can read, interpret, and respond to directly. It is the only language understood by a computer and it can differ depending on the OS on the computer.

2. **Assembly language:** this is any low-level encoding language intended for a particular type of processor. An assembler can be used to convert assembly codes into machine codes.

3. **High-level language:** a programming language that allows a designer to write codes that are independent of a certain computer type.

Programming language is essentially the heart of software. Without it we cannot communicate with systems as they only understand machine code. While human beings only understand high-level languages, machines do not. For the two to communicate there has to be an intermediate. Now, there are many programming languages and we cannot exploit them all. We will only look at some of them that are considered to be in demand. Before then, here are the types of programming languages that you need for better understanding of the languages.

Procedural programming language: one that follows a collection of commands in order. They happen to be some of the common types used by programmers and script. This could be because they employ conditional statements, variables, and functions to design programs through which computers can perform computations and display desired outputs.

Object-oriented programming language (OOP): just as the name goes, this is a model where programs are organized around objects, or data, rather than logic and functions. An object is simply a data field with a unique behavior and attributes. For instance, a person described by unique features like address and name is a good example of an object. OOP is based on four principles; encapsulation, abstraction, inheritance, and polymorphism.

Functional programming language: this is a language meant to influence implementation of software construction by building pure functions. Rather than emphasizing on statement execution, this procedure focuses more on declarations and expressions.

Scripting programming language: a language designed to communicate and combine with other languages. This language is often used alongside Java, C++, and HTML because it is not a fully-fledged programming language. They can get started even with a tiny syntax.

Logic programming language: this is a language that allows programmers to make declarative statements, after which it then gives room for the machine to judge the consequences of those same statements.

An Outline of Some Programming Languages
Python: this is an object-oriented programming language that is built on robust and flexible semantics. It is used by a variety of developers and software engineers, where it is preferred for its ability to quickly integrate systems like glue or scripting language. It is easily read and simple to learn. Python was developed in late 1980s and released in 1991. NASA uses it as a standard scripting language in their integrated planning system. The disadvantage is that it is slower and therefore not scalable for relatively large projects.

Java: equally object-oriented, java is a high-level language that is quite perfect for web development. It's portable across other platforms and does not need memory management, community support, or multiple compilations. Its syntax can get really long now that everything must be an object. Below is a simple code that prints 'Hello World.'

```java
public class Hello {
        public static void main(String []args) {
                System.out.println("Hello World");
        }
}
```

C language: this is a structure-based language used in development of low-level applications. C is very fast and other languages actually inherit syntax from it. Unfortunately, it lacks flexibility of the object-oriented paradigm and does not do anything for you. You have to explicitly build everything and it does not caution when you're about to make a mistake. Below is a simple code for the output 'Hello World.'

```c
#include<stdio.h>

int main(void) {
    printf("Hello World\n");
    return 0;
}
```

C++: this is an object-oriented, universal language, which is also middle-level and an expansion of C. It is an ideal illustration of a hybrid language now that you can code it in C or C++ style. C++ is very fast and possesses modern features, which makes it generally safer than C. The disadvantage is that it is complex to learn and also bears some readability issues. Here is a simple code that outputs 'Hello World.'

```cpp
#include <iostream>

int main()
{
    std::cout << "Hello, world!\n";
    return 0;
}
```

JavaScript: it is a client-side scripting code which operates within a user browser, after which it generates commands. Its commands are processed on a computer and not on a server. It has no relation to java despite its name. In fact, it is usually placed on an ASP or HTML file. JavaScript is employed in web design not only to impact key page components but to also enhance their dynamism. Some of the pages include; scrolling abilities, generating a calendar, and printing date and time.

Which language do you choose?

When it comes to the choice of a programming language, none is perfect. Your preference should be guided by the circumstances surrounding the project. Time is also a critical factor that should guide your choice. For instance, you may decide to develop a basic tower defense game. While you could make the same game in Assembly, it is more prudent to consider Java. Why? When using Java it would take you a month or two to get done, while the same project would take you about a year, not to mention the headaches. If you are in this to find a simple language, you may need to first understand several of them. From there you will be in a better position to make an informed choice even as we look forward to advancing into modern techniques like artificial intelligence and its subsets.

Chapter 4: Differentiating Artificial Intelligence & Machine Learning terms

Machine Learning &Artificial Intelligence systems are among the most popular catchwords in the business world; in many instances, they are put into use interchangeably. There is plenty of confusion when it comes to understanding and differentiating the two. The most common questions include; what is Artificial Intelligence? What is Machine Learning? Are the two related in any way? In most businesses however, marketing tends to ignore their differences for sales and advertising.

Machines and Artificial Intelligence have become an integral part in everyday living. This doesn't mean that the two are well understood. If you're hoping to implement and make use of ML or AI in your business, it's of importance to find out which one you plan to make use of. Artificial Intelligence and Machine Learning are related, but this doesn't mean that they are similar. Choosing either AL or ML can be the determinant in moving your business to the next level.

Artificial Intelligence basically means that machines are equipped to carry out tasks in intelligent ways. These machines are programmed to do various tasks by adjusting to various situations.

Machine Learning can be viewed as an Artificial Intelligence branch; the main difference is that it is more precise than the conclusive concept. The ML concept is founded on the idea that machines that don't require constant human supervision are created. Artificial intelligence has two major subcategories namely applied as well as generalized AI. In this case, the applied AI is pretty more relevant in that it covers various programs including driverless cars in order to import stock trading programs. While on the other side, generalized AI is not that common in practice since it's complicated to create. This type of AI has therefore, the ability to handle various tasks in the same way human would. Machine learning has been explored due to specific breakthroughs in Artificial Intelligence. Therefore, such breakthroughs are specific in assisting humans to handle their tasks pretty more sensibly. Another breakthrough that led to the invention of ML was the internet. The internet has since facilitated extensive storage of information. This has never happened in the past. Machines can also be used in viewing data amounts that have not been accessed over the years in many ways. This is because of the past limitations in storage. As such, the volumes of data formed in excess for human processing makes it easier to perform complex tasks in shorter frames.

The Machine Learning Concept Explained

Machine Learning is an Artificial Intelligence branch which is tasked with providing systems with the ability to improve and learn from experiences automatically; this is achieved without making use of explicit programing. The main focus of ML is the manufacture and improvement of computer programs. These programs can actually use the data they acquire for their own learning.

The learning process starts with data observation i.e. direct instruction, or experience, with a view to looking for data patterns which will facilitate better decision making in future. The main aim is enabling computers to learn automatically and alter actions accordingly, without any assistance or intervention from human beings.

In ML, algorithms procure their skill or knowledge through experience. Machine Learning is an application which is reliant on huge sets of data in reminding the data about finding common patterns.

Artificial Intelligence Defined

Learning in Artificial Intelligence is through acquisition of knowledge as well as exploring ways in which the knowledge can be applied. The goal of AI is to arrive at optimal solutions by increasing the rates of success. Basically, Artificial Intelligence can be defined as an application in which computers make attempts to emulate or mimic things done by humans, in a better manner.

What are the Main Differences between the Machine Learning and Artificial Intelligence Terms?

AI is an abbreviation for Artificial Intelligence; intelligence defines the ability to acquire and make use of knowledge. ML on the other hand stands for Machine Learning which basically refers to acquisition of skill or knowledge.

Artificial Intelligence doesn't aim at increasing accuracy but increasing the chances of success. Machine Learning aims at increasing accuracy, but it doesn't really put emphasis on success.

Artificial Intelligence is a program in co putters that performs smart tasks. The Machine Leaning is a simple concept in which the machine gets to learn from collected data.

The goal of Artificial Intelligence is stimulating natural intelligence to enable it to solve intricate problems. Machine Learning aims at learning from data in regards to certain tasks inn maximizing the performance of the machines.

Artificial Intelligence is basically tool for decision making whereas the Machine Learning system learns new things from previous data.

AI leads in the development of systems that mimics human beings in responding to certain circumstances. Machine Learning is basically tasked with the creation of algorithms that are used for the sole purpose of self-learning.

Artificial Intelligence works by arriving at optimal solutions whereas Machine Learning works by arriving at solutions, whether ideal or not.

AI yields wisdom or intelligence. Machine Learning results in knowledge.

In conclusion, Applications in Machine Learning have the ability to read text as figuring out whether the author is offering congratulations or making complaints. These apps have the ability to listen to pieces of music and to arrive at decisions with regards to if the someone will be sad or happy after listening to the music. There are many possibilities that are provided by Machine Learning apps in connection with neural networks.

Machine Learning makes use of the experiences it handles to source for the learned patterns. Artificial Intelligence makes use of the same experiences in the acquisition of skills or knowledge and how to use the knowledge in handling new environments. Both ML and Ai come with various business applications that are very valuable. However, Machine Learning has become more absorbed in solving serious problems in several businesses. The basic concept in applying the Machine Learning app and Artificial Intelligence system in business has been shown to play key roles in improved overall performance. The major disadvantage however is that making use of AI and Ml in your business automatically translates to loss of employment for manual workers.

CHAPTER 5:
The Most Important Artificial Intelligent (AI) Systems

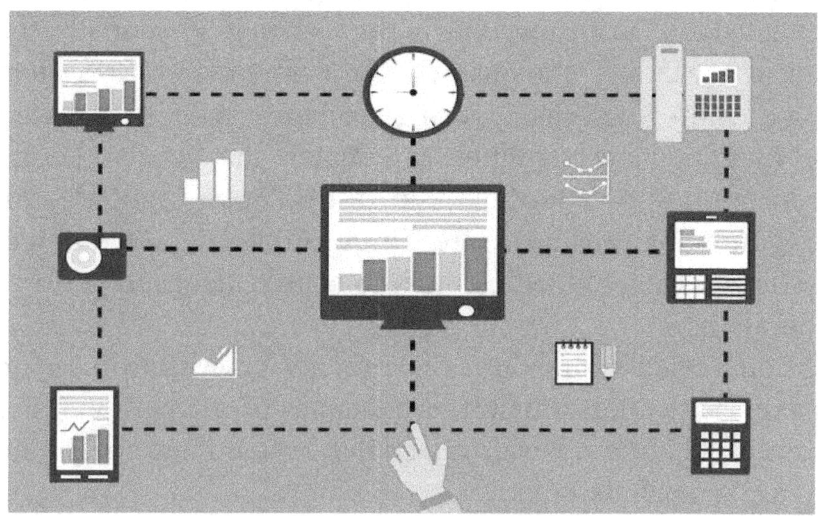

The ultimate aim of developing Artificial Intelligence systems is to replicate abilities possessed by human beings and operate just as intelligently as they do. To do so, developers approach each ability of human beings individually and develop efficient systems for each one. These human abilities include the cognitive skills to think and reason, perceive, learn, identify patterns, make decisions, act, solve problems, plan, remember, and use language skills. In this chapter, we'll look at how AI tries to perform these human abilities.

Artificial Intelligence Systems that act

Artificial Intelligence systems under this category act either humanly or rationally.

When AI systems perform intelligent operations indistinguishable from those of a human being, they are said to act humanly. This measure of intelligence is determined through a test called the Turing test, developed by English computer scientist and mathematician Alan Turin, in the year 1950.

AI systems are said to act rationally if their data manipulation rules sets can simulate what a Turing machine can do. Alan invented the Turing machine in the year 1936. Essentially, the Turing machine is the blueprint for computation. No system can do more than what a Turing machine can do. If, however, its operational capabilities can match that of the Turing machine, the AI system is said to be Turing Complete. All programming languages are Turing Complete. AI Rational systems rely on limited memory to analyze data and perform operations.

Rational agents, or bots as commonly known, are used to maximize the performance or efficiency of an operation. For example, a bot can perform task five or ten times faster than a Turing machine would.

Artificial Intelligence Systems that predict

AI systems that predict are an essential component of Artificial Intelligence for entities and are emerging as critical to the way they plan and set targets.

We use weather forecasts to make decisions on whether to dress lightly or heavily, depending on hot or cold weather forecasts respectively. Similarly, using their predictive models, entities such as businesses and governments use historical data gathered from their subjects to try and predict future outcomes.

Every entity sets goals and targets of what they intend to achieve within a specified period. Using AI's predictive systems, they can get a more detailed foresight of what they will most likely achieve within that period. To make projections of future outcomes, the AI system analyzes relevant data from the subjects and puts all factors that impact on the entity's operations into consideration.

AI predictive systems are not entirely precisely accurate as yet, as AI is still a relatively new concept of computer science. With the future being impossible to forecast all the time authoritatively, this compounds the challenges with using AI's predictive systems. However, they try and give a good impression of what results will most likely be achieved during a specified period. With continuous development and improvement, AI predictive systems will get more and more accurate at predicting future trends and outcomes.

Artificial Intelligence Systems that learn

This form of Artificial Intelligence falls under the category of Generalized AI and is popularly known as Machine Learning. Machine Learning aims to make machines independent and intelligent to learn from numerous kinds of unrelated data sets and apply reason and conclusions unique to each one. To perform this operation, unlike in rational agents, where the systems follow a specific set of instructions, Machine Learning systems rely on inference.

Machine Learning still requires massive development to achieve success as currently this form of Artificial Intelligence is riddled with many failures and is far from being proficient. With vast amounts of data and information available on the internet acting as fodder for Machine Learning systems to learn from, experts are upbeat that they will get better and employable. This increases the chances of success for the systems to produce desired results as purposed — to be independent, intelligent, and employable.

Machine Learning, therefore, requires a significant level of development to start being considered as a reliable commercial tool for entities. Nonetheless, this form of AI has a lot of potential for the future.

Artificial Intelligence Systems that create

Two Machine Learning systems are set against each another to make creations using AI successfully. One system creates an object, such an image, and the other works to find fault with the invention, thus making the two systems co-work autonomously. The concept of connecting two or more AI systems to work together to perform a task is what is known as Artificial Neural Networks, ANN.

An AI system that invents an object would first have to learn from similar data sets to replicate the creations, making its own. This operation is made possible using a new concept of Machine Learning called Generative Adversarial Network, GAN.

GAN has aided AI systems to make creations successfully. More recently, tech labs such as DataGrid and Samsung have successfully created both static and moving images using this method. Known as deep fake, AI systems can be able to generate very real-looking motion images from a single static image. Samsung AI Center was able to successfully use the famous portrait of Mona Lisa to make a video from it. Even more intriguingly, it was lip-synced to appear to speak.

As mind-blowing as this concept is, it poses a serious concern as people with bad intentions may use the technology for malicious purposes. Therefore, there needs to be tight regulation as to who can access this tool.

Artificial Intelligence Systems that relate

These systems learn the subject's location, habits, and preferences on their platform and relate the most relevant items the user is most likely to consume. The aim is to filter out unnecessary items and suggest the ones that would be most relatable with the user's data.

This tool of Artificial Intelligence is mostly used by search engines, social media sites, media content websites, and e-commerce sites to advertise products and suggest content relating to individual users on their platforms. It has proved to be an essential tool for driving up sales for e-commerce websites and increasing consumption within social media sites and media content websites.

Artificial Intelligence Systems that master

Developers aim to create a single computer system that can perform all intelligent tasks, unlike anything seen as yet in the Artificial Intelligence scene. It would arguably be an invaluable invention. Theoretically, it could combine all human beings' cognitive abilities already replicated in Artificial Intelligence functions and those still under development and perform all of them efficiently.

Given the superior power and abilities such a machine would possess, people have concerns and fears about it — and rightly so. Humans rule over creation because their abilities are superior to the rest of creation. If a single machine possesses all the skills of humans and more, then, these super intelligent machines would be superior to humans. They would have the power to rule over them, meaning either the extinction or insubordination of the human race. This chilling depiction is the picture of robots mostly portrayed in movies.

However, proponents of Master AI argue for the positives of these systems, such as having a single robot able to provide treatments to ailments and create a sustainable way of conserving the climate and enhance the security of citizens. Question is: "Can humans trust and tame a creation whose intelligence is significantly superior to theirs?" It would take a long time to find the answer to this question as the realization of this concept is considered presently unachievable.

Artificial Intelligence systems that evolve

In nature, the process of evolution aims at creating an improved form of a human being or organism to adapt better to its changing environment. Similarly, in the evolution of AI systems, a Machine Learning algorithm works in a topology of Artificial Neural Networks to create a more efficient and accurate system.

For over three decades, the most common method used in the evolution of Machine Learning algorithms is Gradient Descent, GD. GD identifies Neural Networks with the lowest cost functions. The lower the cost function of a system, the more efficient the system is at achieving the desired results. Therefore, Neural Networks learn from only the best algorithms and optimize to create a better one. To get best results projections, the number of Neural Networks has to be just right — not too few, not too many.

CHAPTER 6: REGULATION & ETHICS OF ARTIFICIAL INTELLIGENCE COMPANIES: SAFETY, LAW, AND GENERAL DATA PROTECTION

Artificial intelligence has become a phenomenon in the present times because of the many benefits associated with it.In the business world, there are both benefits and risks that are associated with Artificial Intelligence. Businesses are making use of AI programs with the aim of improving their business results, profitability and productivity. The application of solutions driven by Machine Learning and Artificial Intelligence applications has become a very common practise. However, this doesn't mean that it is all smooth sailing for businesses using Artificial Intelligence.

There are several risks of artificial intelligence including:

Data Availability
During the phase of Ai implementation, data is often of low quality and inconsistent which makes it difficult to get value from AI.

Skills Shortage
There is the skills shortage problem for businesses trying to implement AI; there could be shortage of technical staff who are experienced. This could translate to extra costs to train employees for the successful operation of AI.

Implementing Artificial Intelligence is costly
Regular updating of software programs is needed to keep AI machines running smoothly. In cases where there is staff shortage, businesses are forced to outsource which is costly. The additional costs incurred in data training models can prove to be very costly especially for small businesses.

Other limitations related to Artificial Intelligence include:
There may be delays in the implementation process depending on the Artificial Intelligence programs being implemented
Misunderstanding the Artificial Intelligence programs which results in integration setbacks
Artificial Intelligence applications may not work well if interacted with other programs
These risks have necessitated the drafting and subsequent implementation of safety, law, general data protection regulation and ethics of artificial Intelligence companies. A framework tasked with the management of Artificial

Intelligence ethics has been developed due to the significant risk potential associated with the same.

What are some of the core principles that guide governments, industries and developers in the ethical deployment of Artificial Intelligence driven systems?

Generating net benefits

It's a must for Artificial Intelligence systems to work by generating benefits for users which overshadow the risks.

Causing no harm

Artificial Intelligence must be designed in such a way that people won't be deceived or harmed and negative outcomes must also be minimised.

Legal and regulatory compliance

The systems installed with Artificial Intelligence must adhere to all government obligations, regulations and laws.

Protection of the users' privacy

Artificial Intelligence systems are compelled to see to it that data is kept confidential and protected. Harmful breaches in data must be prevented when using the AI systems.

Fairness

Individuals, businesses or communities using Artificial Intelligence systems must not be subjected to unfair discrimination. These systems should be implemented in a manner that they are won't cause unfairness which results from training biases.

Explainability and Transparency

Users must be well informed about how the AI algorithms that are being used affects them. Users must have proper understanding about the data or information that these algorithms use in making decisions.

Contestability

A competent process that can challenge the output or usage of algorithms must be in place since these affects users directly.

Accountability

Organisations and the people responsible for the development and implementation of the Artificial Intelligence algorithms must be accountable and identifiable for the resulting algorithm impact. This applies irrespective of whether the said impacts are intended or unintended.

How does the General Data Protection Regulation (GDPR) Safeguard Users in the Artificial Intelligence Centred World?

1) GDPR in privacy protection

The General Data Protection Regulation causes a hurdle in the implementation of Artificial Intelligence. At the same time, it is an important factor when it comes to the privacy of the users. Trust with regards to data privacy has been eroded as a result of data breaches. The good thing about this risk in Artificial Intelligence is that businesses are in a position to restore the lost trust simply by observing GDPR. This means that the companies take a straight forward approach that focuses in Artificial Intelligence privacy.

GDPR demands that holding data for longer periods than necessary isn't acceptable. This raises questions as to whether Artificial Intelligence is being silenced. This can be quite confusing since AI relies heavily on past data to process information that enables the system to make decisions.

2) GDPR in automated profiling and decision making

There are specific provisions of GDPR that target decisions that are based on Artificial Intelligence, specifically with regards to automated profiling and decision making. Set provisions must be adhered to strictly to ensure that the users do not get biased.

3) GDRP in information gathering

Ensuring that the company clearly understands the data that is collected and the manner in which the data is processed is the first as far GDPR is concerned. Documentation of the data type collected, the source and the channels via which it has been collected is crucial.

4) GDRP in risk assessment

Prior to making use of automated decision making, a Data Protection Impact Assessment (DPIA) must be carried out. DPIA allows the organization to gauge the risks that come with automated decision making. DPIA aims at examining risks to subjects of data at every data processing step.

5) GDRP in the management of third parties

It's important to have an understanding of the privacy and security controls that are used by your vendor. Be sure to find out from your vendor whether he has all the required industry certifications; this is very important in the verification of vendor assessments.

In conclusion, Artificial Intelligence and General Data Protection Regulation emphasizes that the big data that is used in Artificial Intelligence is directly opposed to purposeful limitation of data retention and collection. It's quite clear that as far the safety for AI users is concerned; following laid out regulations helps in safeguarding privacy of the users.

Be sure to find out prior to AI implementation that your vendors observe these guidelines to the letter. This may seem like a daunting task but the good thing about it is that users feel safer and more confident in transacting in your company when they have the guarantee that they are safe.

Chapter 7:
ARTIFICIAL INTELLIGENCE IMPACT IN BUSINESS ASPECTS: OPPORTUNITIES, BENEFITS AND RISKS

We come across the term Artificial Intelligence from time to time as we get along with our daily lives. There's more to Artificial Intelligence than 'Alexa' or 'Siri' which are great influencers in making our activities in the business world easier. The target of Artificial Intelligence is making computers to be smart enough to mimic the behaviour of the human mind. This basically means that Artificial Intelligence greatly helps us to achieve our business goals.

In the present age, Artificial Intelligence applications are in use constantly. You may have used the Ok Google feature in many instances but probably never gave it much thought. This

Google feature is an application that make everyday living easier for human beings.

Artificial Intelligence in business is a concept that is very interesting and fascinating. Technology in the present day is picking at a very fast pace meaning that businesses have to keep up with this pace to ensure that they utilize technological advancements fully to their advantage. It's important to note that AI is being used in the improvement and transformation of businesses.

The Artificial Intelligence field is rapidly advancing. The AL applications enabled major breakthroughs in game playing, autonomous robotics, speech and image recognition. Mostly, AI promises many benefits; better and cheaper services and goods, medical advances and fresh scientific discoveries. At the same time, AI raises come concerns; security and safety, inequality, bias and privacy among others.

Most of the AI programs in use today are very narrow; meaning that in some instances they may be incapacitated when it comes to handling certain tasks. The AI applications aren't necessary superior to human beings in all aspects; they could be having better performance than humans in one aspect or another but not all.

Where are Artificial Intelligent applications applied in Business?

The applications in AI are presently in use in different areas in the business field including:

- Business and sales forecasting
- Voice-text features
- Spam filters
- Security surveillance
- Automated insights

- Online support for clients over the internet and automated responders
- Process automation

It's very clear that the power that has been bestowed upon machines has helped greatly in solving business problems, but, at the same time it has led in destruction or risks in some instances. The following are some of the benefits/ opportunities and risks of using Artificial Intelligence in business:

Advantages of Artificial Intelligence in business discussed:
Reduction of the rate at which errors are made in business
Decision making by machines that are installed with Artificial Intelligence applications are purely base on prior data records and algorithms, greatly reducing the chances of errors occurring. This is a great accomplishment, as complex problems that need to be solved in businesses is done without any room for error.

Top business organizations are making use of digital assistants in interacting with customers. This accelerates the rate at which feedback is given to clients, providing the best quality assistance to users.

Making informed decisions
Artificial Intelligence applications have no emotions hence making them better placed to make unbiased decisions, in a short period of time. A good example is in the healthcare facilities; AI programs have facilitated provision of efficient treatments mainly by bringing down the risk of misdiagnosis.

Ability to perform tasks continuously
Machines installed with Artificial Intelligence programs can work for long periods of time without getting tired. This is a huge advantage in business as compared to human beings in regards to productivity. External factors never affect the efficacy of machines which is a very important benefit in business.

Disadvantages of Artificial Intelligence in business discussed:

Artificial Intelligence is very costly to implement and use
The total cost of installing, repairing and keeping AI machines in good shape is very high. Businesses with huge savings can implement the AI programs. However, companies that don't have sufficient funds will not be in a position to make use of these applications in their strategies and processes. This means that these business get locked out from enjoying the Artificial Intelligence benefits.

Displacement of manual workers
This is the main disadvantage of Artificial Intelligence; these apps have led to people being displaced from their places of work. This happens where a single machine gets incorporated in a business and performs a task that was being handled by several people, at a go. This automatically means that some employees are rendered jobless.

Over reliance on Artificial Intelligence Machines
Human beings are becoming increasingly dependent on AI machines; it's slowly becoming very hard for human beings to accomplish several tasks without the help of machines. This

has led in decreased thinking and mental capabilities in humans.

Machines using Artificial Intelligence can only perform specified tasks

AI machines have already been programmed to perform only outlined tasks. These machines can't be relied upon to take to unfamiliar environments or make decisions outside the areas that they are programmed to handle.

How will Artificial Intelligence Impact Businesses in Future?

The future in the business world will greatly be determined by the use of various Apps in the simplification of various tasks. The Artificial Intelligence is most definitely among these Apps that are bound to play a key role in business running in future. Machine Learning refers to the art of applying and designing certain algorithms that have the ability of learning events based on past occurrences. For example, complicated algorithms have the ability to analyse past cases of frauds and managing similar happenings beforehand. The only hitch in Machine Learning is that inn instances where there aren't any previous cases, then these algorithms can't work. Always remember that ML and AL have a number of differences but always work better when incorporated together.

By making use of Machine Learning and Artificial Intelligence applications in your business, the systems won't have limitations concerning the instructions that had been given out by the programmer. This means that the programs will not require plenty of programming compared to ancient programs. The Artificial Intelligence app is bound to alter the manner in which businesses will be running in the coming days. Some of the changes that you can expect in future in business as a result of using Artificial Intelligence include:

Finding better candidates in a quicker and easier manner for your business.

In the work place environment, we depend a lot upon machines to do or complete certain tasks for us. Artificial Intelligence will be used in the identification of certain factors i.e. the period of time an employee is positioned in a specific place, the cost of hiring an employee, specific job positions that tend to be filled more quickly as opposed to positions which take much longer to be filled, etc. This will help you in understanding the behavioural trends of your employees.

There may be the increased ability in the identification of any hindrances while carrying out the recruitment process; this can even assist you in cost reduction in this process. This also ensures that you get to hire the most suitable candidates in the shortest time possible.

Saying goodbye to theft and errors

It's very normal for human beings to make mistakes. No matter how minute these errors appear to be, they end up affecting your business in one way or another. There is a great concern that in the coming days, machines or robots with Artificial Intelligence programs will eventually render people jobless. These machines are quickly becoming popular in businesses since they increase business efficiency greatly.

If the AI system is developed properly, then human errors can be done away with completely, creating an environment that is almost risk-free for the business.

Companies that depend heavily on smartphones, laptops and the internet are always prone to theft or cyber-attack. Once AI familiarizes itself with the decodes and system deviations, algorithms and patterns, it reveals malfunctions and attacks in process, which prevents the business from incurring possible massive losses.

Simplifying and Upgrading the customer service process

This is among the most critical aspect in any business. It can be a task that is also very time consuming. You can imagine taking several hours to solve a single complaint from a customer. But with the trend in which customers raise their complaints via chatbots, the time consumed in resolving the same is really reduced.

Although there is no personal touch when using chatbots, with time, they will. Tools such as the Digital Genius have been implemented in creating human-like conversations that are flawless with customers. With the aid of Artificial Intelligence programs, companies are better placed to increase their data collection capacity and efficacy. This data is then processed which helps business owners to know the behaviour patterns of the customers. This will be vital in updating the policies of the business.

Measurement of Big Data

AI and Big Data are terms that will be key players in the future for the successful running of businesses. Artificial Intelligence businesses will not be in a position to work in the absence of Big Data in future.

It's important for business owners to ensure proper data preservation in vast amounts, which will ensure that no information is lost. Machines that have the capability of carrying out content analysis and reasoning that is evidence-based can improve the making of decisions for businesses edged towards people's management.

Branches of Artificial Intelligence I.e. Deep Learning and Machine Learning can be used in the management and analyzation of big data. It's basically used in sourcing for predictions, patterns and trends. This will be of great impact and importance in a finance business or stock market.

Lead generation;
This is defined as a process involving the cultivation and identification of potential customers for any business services or products. Expenses that are directly linked to the acquisition of customers through lead generation are considered to be very important. Artificial Intelligence businesses are making use of certain AI programs that enable better data processing as compared to human beings. These programs analyse data from social media platforms which enables business owners to have a better understanding of the interests of their customers.

In conclusion, the above mentioned business areas can be grasped better with assistance from Artificial Intelligence programs. Artificial Intelligence also assists in several areas in the business world such as time management and administration and many others. This shows clearly that Artificial Intelligence is undoubtedly going to become the key feature in business, in future. It would be a great idea to adapt and start making use of Artificial Intelligence for better and greater success.

Chapter 8:
Data Science and Data analytics in Artificial intelligence

Nowadays, it is utterly impossible not to come across analytics and data science mentions in modern online or physical environments. In short, analytics and data science or combinations of the duo are some of the terms that are trending in our present world. Consequently, various foundations now offer jobs based on full conversance to data science and analytics. Therefore, it is an understatement not to privilege the substantial data sciences and analytics' 'tech' invasion in corporate and academic environments after the swift evolution of the internet and digital economies. The intelligent Data Science and Analytics technologies are gaining adaptation in industries, individuals and business establishments.

Data Science and Analytics involves the gathering, grouping and examination of data and connote its outlines, facts and aptitude. Concurrently, Data Science and Analytics can be classified on the base of their descriptive, predictive and prescriptive aspects. However, these aspects need to be theoretically, technologically and methodically developed to facilitate the growing global demands.

Innovation in the Data Science and Analytics Environment

Although certain basic constituents of Data Science and Analytics have long been in existence, their presence has overseen tremendous, innovative and opportunistic advancement in Machine Intelligence. Analytics as services are relatively new additions to the Artificial Intelligence world. In the accord of the involvement of analytical systems in data comprehensions, rapid migration of fiscal and social deals in the online world has been perceived. This has, in turn, facilitated the digital apprehension of vast statistics. Moreover, there has been a significant development in the contemplation of human exposition configurations and expositions. Consequently, these developments have stretched out the scope and accessibility of statistic sets. Accordingly, an exponential boom of inquiry expositions extending to various ranges of investigations has been observed. Also, boosted by the innovation of 'smarter' machines, the nature of inquiry has also improved. The new 'smart' machine features have been brought by, by the elevation and concurrent instrumentation of algorithms into inherent interactions with the algorithms. In addition to the collection of data for exclusive human analysis or vital record-keeping, a more reasonable capture of data backlogs, entirely, for un-envisioned hypotheses has been introduced. With the characteristic of optimal gather and understanding of data, computers have evolved to 'intelligent' probing machines. Computers, through refined probing, can consequently create and discover new knowledge that would be, long due in reliance to human hypotheses (Agarwal and Dhar,). This proves and justifies a probability of theories generated by computers.

However, with the evolution of algorithms come elevated probing desires. These desires are key building blocks in the improvement of algorithms and structures. Consequent to the increased, saturated and unstructured data, further evolution of algorithms is evident further focused on heterogeneous and uneven data. The research frontier is now involved in the creation of cutting edge developments in the areas of such as; transcript and image processing, and natural language processing, consequent to the uncertain existence of any or ultimate data incorporation and analysis limits transversely present on the internet.

In support of existing networks concerning humans and product ranges, a requirement of better sampling and validating algorithms are needed in the discourse of these 'societal' features (Ugander et al. 2013). These new developments involving interactive and encompassed algorithms evolve from a range of subjects including computer science and IS or any other disciplines. In concurrence to Kramer et al. (2014), research proves the manipulation of internet users' moods through social-based algorithms. Casual networks can be derived from variables through experimentation of people. However, conducting these experiments without informed approval may implicate the research with unethical privacy conduct. Nevertheless, proof of medium for the development of theories based on the investigation of social and other disciplines is evident.

Data science and Data analytics on Reasonable Benefit for Information Systems

As accorded by Aral and Walker (2009), this is arguably a golden era for Information Systems research. As evident from information incorporated in some commonly visited scientific channels such as PNAS, Data Science and Data Analytics research is routed in Information Systems (Aral and Walker, 2012). Consequently, common questions of the reasons for our choices and reasons for interest in Data Science and Analytics have risen. What influences the choices we make in social networks... sway or homophily? Accordingly, this and other questions offer scientists opportunities to discern, inquire, and connote social behaviors on an inclusive scale.

Data science and Data analytics on Investigation Questions

As coined by Lucas et al. (2013), following the Information Technology revolution, Data science, and Data analytics sanction the creation and implementation of studies. Moreover, they facilitate extends to existing research under explored outlines such as; learning result mediations, and satisfaction, online tutoring, and additional information of exceptional gauges. Furthermore, Data science and Data analytics facilitate the decoding of a network; distinct, organizational, and societal existences and their connoted outcomes.

Data science and Data analytics, Predictions and Explanations

As proven by Weil (2014), it is justifiable to acknowledge the importance of Data Science and Analytics in the providence of remarkable randomization of research from unrealistically massive observational data. These operations are applicable if data and aspects of data can be satisfyingly and respectively explained or predicted. In this accord, Data Science and Analytics offer researchers potentials for innovative discoveries.

Through Data Science and Analytics, scholars, researchers, and everybody have the advantage of creating new scientific prospects. As coined in the above Data Science and Analytics integrations, it is a superb and exciting period for research to evolve beyond Information Systems and general 'science'. Concurrently, the delivery of data in the most accurate and easy forms are possible. In these accords, growth in education, technology, societies, businesses and other aspects of human existence is inevitable. Data science and analytics offer managerial prepositions to businesses since they are integrated with technological systems i.e. Artificial Intelligence and thus help improve decisions in businesses. With Artificial Intelligence abilities to plan, organize, lead and control, the management of the businesses is made easy through innovation, prediction, and investigation.

CHAPTER 9:
ARTIFICIAL INTELLIGENCE APPLICATIONS IN FUNDAMENTAL SECTORS

No doubt artificial intelligence is at the edge of transforming every section of our economy. It is exciting to see the many applications that AI accommodates in today's society, and more so the significant changes that it is making in those sectors. Allowing computers to make human-like decisions helps a lot in automation of routine tasks in businesses. The system learns from experience such that it is able to understand every customer, examine their behaviour, personalize experiences, and significantly cut back on operational costs. In fact, artificial intelligence is likely to become an indispensable tool in all medical disciplines. As we seek to understand how best AI disrupts and shapes different industries, we will first look at its application in the healthcare sector.

Healthcare and Medical Diagnosis

As AI continues to infiltrate diverse aspects of our lives, the impact is quite perceptible in the healthcare industry. Although there is already a lot to show for it, there is still wide-reaching potential in artificial intelligence in healthcare. Do you know that drug discovery and mobile coaching remedies are under the umbrella of things being achieved through machine learning? Well, let us get to the details on the much that we know AI has done in this industry.

Faster development of drugs

There are AI-powered firms that are set to collaborate with giants in the pharmaceutical industry so that they can significantly augment the efficacy of drug development. Undoubtedly, artificial intelligence has remarkable computational power considering that it employs personified knowledge gained from solutions. Honestly speaking, the pharmaceutical industry is experiencing major challenges in upholding their drug manufacturing programs.

Reduced efficiency and increased costs of research and development are just some of the major setbacks making the entire process notoriously expensive. The effectiveness of most of the analytical practices in drug development can be enhanced through AI and machine learning. Drug development incorporates four main stages which include; identifying target molecules for intervention, discovering effective drugs, expediting clinical trials, and finding biomarkers for diagnostics. Amazingly, artificial intelligence has already been employed in the four mentioned stages and was successful.

Identification of target molecules for intervention: this initial stage is seeks to understand the pathways, or the biological source of a disease and also its resistance ability. From there, you then have to discover positive targets for combating the disease (usually proteins). Discovery of viable target molecules requires a lot of data. Thankfully, the extensive availability of high throughput systems such as deep sequencing has made it easy to access more data. Even so, integration of an array of data sources with the traditional methods remains a challenge, and this is where artificial intelligence now comes in.

Discovery of effective drug candidates: upon identifying a target molecule, you need to now get a compound with the ability to interact with it (the identified molecule). The process entails screening of thousands or millions of probable compounds to gauge their affinity as well as toxicity. The compounds in question could be artificial, natural, or bioengineered. Unfortunately, the software in use at present is unreliable as it is frequently erroneous and has lots of false positives. As such, it always takes unusually long to identify the optimal drug candidates, usually known as leads. Now, with incorporation of AI, it is easier to have the ML algorithms learn how to forecast the correctness of a molecule. This can be achieved by incorporating molecular descriptors and structural fingerprints, a process that saves a lot of time.

Accelerating clinical trials: the process of finding appropriate candidates for experimental trials is quite difficult. Unfortunately, selection of the incorrect candidates ends up lengthening the trial, which eventually costs a lot of resources and time. Machine learning can effortlessly expedite the clinical trials design. In this case, ML can be used to automatically identify fitting candidates. At the same time, the ML algorithms can help in ensuring proper distribution for clusters of trial participants. The best thing about algorithms is that they can detect an unproductive clinical trial from the onset and send an early warning, giving room for the researchers to chip in and save the development process.

Find biomarkers for diagnostics: it is only possible to offer treatment to patients when you are guaranteed of their diagnosis. Some of the techniques used to diagnose the diseases do not only require high expertise, but are also very expensive. Whole genome sequencing is one such method that also requires use of sophisticated lab equipment. Biomarkers are molecules contained in human blood. These are the particles that ascertain whether or not a patient suffers from a given disease. Biomarkers, therefore, are quite effective when it comes to diagnosis, making the entire process cheap and secure.

These molecules can also be used to know the progression of a disease, which makes it easy for the medics to monitor the effectiveness of the drug. But again, it is also very expensive to determine the appropriate biomarkers for a particular disease. You'd have to screen hundreds of thousands of probable molecule candidates. At this point, AI can be used to automate a significant segment of manual work and classify the bad and good candidates, hence aiding the doctors in focusing on analysis of the best prospects. Biomarkers can be used in; detecting a disease at the earliest possible stage, identification of the risk of someone developing a disease, any likelihood on progression of some disease, as well as monitoring the response of the drug on a patient.

Diagnosis of diseases

Correct diagnosis of ailments takes years of arduousness and time-consuming processes. Diagnostics in most disciplines usually demand a number of experts far exceeding the available supply. As a result, this strains the available doctors while at the same time it impedes life-saving diagnostics. Thankfully, ML, mainly deep learning, has made significant advances in automatic diagnostics, making it economical and more accessible. Machine learning algorithms are smart such that they can learn to see patterns just like the doctors do. However, they need thousands of neatly digitized and concrete examples as they cannot read between the lines in study materials.

Please note that ML algorithms are particularly helpful in fields where the diagnostic data examined by a doctor is digitized. Such areas include, but are not limited to; detecting strokes or lung cancer by use of CT scans, evaluating the risk of cardiac arrest using electrocardiograms and cardiac magnetic resonance imaging (MRI), categorizing membrane lacerations in skin images, and identifying diabetes complications of the eye from the relevant images. The algorithms have become smart in diagnostics especially with availability of plentiful good data in the said cases. The good thing about them is that they are naturally data-heavy, and hence ideal for AI application. AI is not replacing the medics any time in the near future. It is more of a collaborative affair.

Personalize treatment

To increase the lifespan of a patient, treatment has to be tailored. As it is, different patients have different responses towards treatment schedules and drugs. However, identifying the particular factors that affect the treatment option is quite intricate. Machine learning has the capacity to automate this complex statistical task. The ML algorithms can learn how to effectively determine the characteristics depicting that a given patient will have a particular reaction towards some given medication. Precisely, there are algorithms that can help in projecting a patient's potential reaction towards a certain treatment. Usually, the system takes advantage such that it learns from cross-referencing comparable patients where it matches up their medications and results. Doctors rely on the resultant outcomes to design a reliable treatment plan.

AI for Real World:
Positive and Negative Aspects for Clients and Consumers
One undeniable fact is that AI-driven results have already had a massive impact on customer experience, and they will still continue to do so. It has reached a point where the clients and consumers acquire unprecedented information about a product before deciding whether or not to opt for it. Alternately, businesses have also benefited as they can use churn data to foretell the clients who're likely to leave or stay. Even with all the good, we cannot refute that there will come along some negative consequences as well. Discussed below are both sides of artificial intelligence for real world.

Positive Aspects of AI for Clients and Consumers
Customer satisfaction: this is the hallmark to delivering a first-class client experience. When coded properly, artificial intelligence has low fault rate as compared to humans. You realize that most client interactions require human involvement including social medial exchanges, phone calls, online chats, and even email. AI has made it possible for the computers to handle client enquiries with incredible accuracy, precision and speed. This is especially because when combined with ML, it becomes even better for the platforms

to communicate with the customers, making their experience remarkable.

Simplified item search: if you have not known, then it is about time you appreciated that AI has already fortified its authority in the digital era. It is very likely that you interact with it on a daily basis, even though in its subtlest and most modest shape. In your phone or computer operations, you might have noted that there are ads that keep popping. Now, have you ever taken time to think how those ads always happen to be suited to your taste? Artificial intelligence monitors consumer behavior on social media and then goes on to influence their choice.

Intelligent customer service: it is very possible for artificial intelligence to gain essential client insight. This is where AI-powered computerized associates can learn a client's behavior and help them with selection by recommending products inclined towards their fit and needs. It builds a customer's confidence when they get a wide array of products that seem to match their needs and preferences.

Predicting outcomes: artificial intelligence is incredible in the sense that prediction of outcomes is possible by use of data analysis. For instance, when you have stock, the system can forecast the possibility of your commodities selling, and the margin by which it will happen. At the same time, it can also foretell when demand for your commodities will lower. As an entrepreneur, you need this fundamental information to determine the products to stock and the suitable amounts to source during a particular season.

Monitoring conversion rates: there are those traditional methods that people in the traditional days turned to for conversion. These techniques are no longer very effective especially for a modern business. When a potential client visits your site, you need to have a reliable way of turning them into customers. The said traditional method can even take several months, which means losing business. If implemented properly, AI can track all the essential customer records and draw a conclusion on the possibility of a visitor becoming a client. By so doing, the system helps you to save precious time that would otherwise be used to sleuth down inappropriate metric.

Negative Aspects of AI for Clients and Consumers

Job loss in some particular fields: while it is true that artificial intelligence will create many jobs, it is also likely that there are some that will be lost. There are those tasks performed by human beings that will soon be taken over by the machines. It means there will have to be a lot of training done so that our future workforce gets fully equipped. Of course there should be education programs that will need to be initiated to assist the current workforce to prepare for transition into new offices that will make use of their distinctive human abilities.

AI bias: remember that AI algorithms are created by human beings. As such, it is likely for them to contain in-built bias. This could be caused either inadvertently or intentionally by the persons introducing them into the system. Should the algorithms be bias or they get introduced to biased datasets, then it means they will automatically produce biased results. The results could cause unintended consequences just like it happened with the Twitter chatbot by Microsoft that depicted a lot of racism.

Accelerated hacking: even though there is great acceptance by the consumers and business owners to roll out AI-powered services, there is also the fear of privacy and data protection. After all, AI is good at expediting results in diverse instances, making it difficult for humans to follow along. That is why malicious acts like instilling viruses into software programs and phishing have become common.

Global regulations: this is very necessary for safe and effectual interactions globally. Regulation can only be done with introduction of artificial intelligence laws and regulations amongst governments. Since technology has made the world a global village, it means that we are not isolated from each any more. As such, the decision that a country makes regarding AI is likely to affect other countries, whether positively or negatively.

Artificial intelligence terrorism: technology is open for all and this means there could also be forms of terrorism depending on AI. Think of development of autonomous drones and robotic swarms. Our defense organizations will need to fine-tune all the potential threats. It will take extensive expertise and human reasoning to ensure minimal adverse impacts.

Artificial Intelligence for Banking: Critical Banking Functions Most Likely to be Transformed by AI

The financial industry, being data-weighty, has also proved that indeed artificial intelligence is able to change its landscape as well. Amazingly, this is happening even in the fields considered to be traditionally conservative. Without doubt, the banking sector is a fertile ground for artificial intelligence. If deployed properly in the next few years, AI will certainly boost revenues and potentially minimize costs. At the same time, there are high chances of a rapid increase in account and transactional security. This is especially with the expansion of cryptocurrency and blockchains. Looking at it critically, you also realize that when the intermediaries get reduced or eliminated, transaction fees will also be toned down.

With cognitive computing, all the applications and types of digital associates will continue to perfect themselves. Managing personal finances will only get exponentially easier as smart machines will do much of the work. Even tasks like organizing tax filings and execution of some of the long and short-term jobs will be done effectively by these machines. This will bring along an entirely new height of transparency, which is based on know-your-customer reporting. The intense due diligence checks performed will also be more effective and will replace the many hours that human beings take.

Even so, banks must first have a clear strategy for them to get the transformation right. There are many fields that are in the process of automation, but this might not be as easy as said. Let us look into some of the crucial functions that are likely to be transformed by AI. In fact, all of these areas have been impacted by AI, only that improvement needs to come in handy for some of them.

Risk management: data handling is quite crucial in the finance sector. AI has huge processing power, which when combined with cognitive computing assists in managing structured and unstructured statistics. Unlike humans, artificial intelligence evaluates the record of previous risks and identifies early signs of potential future problems.

Fraud prevention: battling fraud in the financial sector is getting easier as AI helps in prevention of credit card fraud, which has been the most prevalent. The algorithms can learn a customer's behavior and analyze their buying habits. In case of an uncertain pattern, the system raises an alarm. These days the banks are also employing artificial intelligence to reveal and combat money laundering.

Personalized banking: AI powers chatbots that are known to offer comprehensive solutions to clients, hence reducing workload at the call centers. There are also smart tech-powered virtual assistants that are voice-controlled and they are gaining traction quite fast. Amazon's Alexa is a perfect example and it keeps on getting smarter by day, which means there are more tremendous improvements yet to be experienced. Some of the big banks in US are already operating with applications through which their clients can interact with customer care for assistance, get reminders to settle bills, and also plan their expenses.

Credit decisions: financial institutions suffer when borrowers keep defaulting. For this reason, it is crucial for these lenders to evaluate a borrower's credit score before deciding whether or not to lend them money. Artificial intelligence provides a more precise evaluation of a borrower. Usually, it incorporates a wide array of factors, which helps a lot in coming up with a data-backed choice that is also well informed. The algorithms assist in distinguishing potential borrowers with a high risk of defaulting from the creditworthy ones who may be lacking an extensive borrowing history.

Process automation: robotic automation of processes is a sure way of boosting productivity and cutting operational costs. The industry leaders focus on such aspects like intelligent character identification for obvious reasons. Such aspects computerize a range of mundane and lengthy tasks that only blow up payrolls in the business setting. Normally, the AI-enabled program validates data and creates reports based on a variety of parameters. They also analyze documents and extract more information from necessary forms such as agreements and applications.

In essence, we cannot overemphasize the benefits of artificial intelligence in financial services; they are too many to overlook. According to a report given by Forbes, a significant percentage of senior financial administration is anticipating positive transformation from deployment of AI in financial sectors. This gives us confidence that indeed banks can effectively lead the shift to artificial intelligence.

Although only a relatively small fraction of financial organizations has embraced implementation of AI in their processes, there is a lot of hope. Certainly there is fear that the endeavor will take a lot of time and consume a lot of finances, and this is normal. Nonetheless, this is a technological wave on the move that they can't shy away from forever. If anything, holding up from embracing it now may only cost a lot more in the long run.

CHAPTER 10:
Growth of business value from Artificial Intelligence

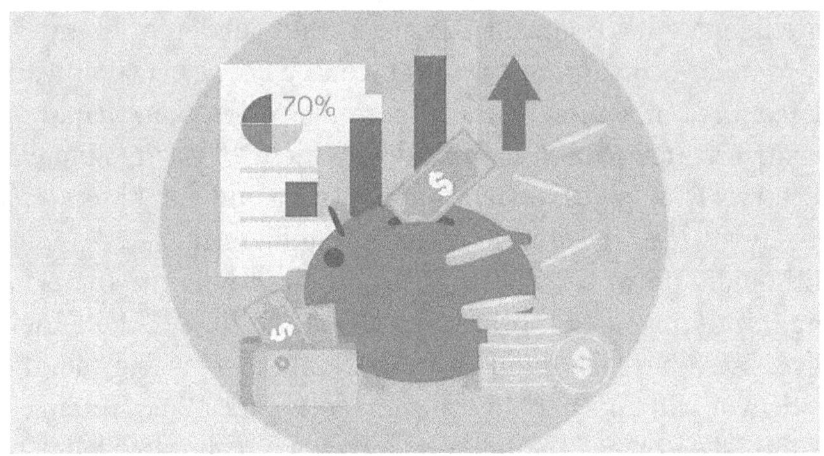

As defined by its consequents including abilities to sense, understand and perform independently, Artificial Intelligence has attracted executives, scientists, and researchers. As a result of frequent media attention, desires have grown in capitalists to exploit the innovative business solution Artificial Intelligence assures. Accordingly, companies involved in technology have grown interested in the exploitation of AI as a venture (Bataller, C. and Harris, J., 2018). Well-contoured acquisitions such as Facebook, Google, and Apple have developed desires in Artificial Intelligence expertise precise to robotics, communication gestures and computer visualization among other fields.

Consequently, the demand for Information Systems has vividly amplified. This has been influenced by devoted desires by companies in the exploitation of these fields. However, with any new promising prospects, business executives are faced with the dilemma of differentiating hype from potential offered by Artificial Intelligence. Furthermore, as a result of skepticism and ignorance of fiscal values consequent to these technologies, Artificial Intelligence prospects face tough gusts. Nonetheless, executives should keep in accord the benefits of AI, rather than dwell in the uncertainties. In these accords, executives should appreciate the diversity and abundance of the AI field.

Artificial Intelligence: Independent, Discerning, Understanding and Acting Systems

With AI technologies, computers are independently capable to perceive, understand and act on data collected. Moreover, AI systems can learn from familiarity and adjust accordingly to the knowledge acquired.

Ability to Sense

Considering the optimization of business habituation, visual technologies i.e. facial recognition may be used to recognize; customer characteristics, and security liabilities among others. Moreover, Video analytics, an additional sensing technology may be used to cement the recognized aspects as mentioned above.

Ability to Understand

AI systems also can understand using technologies in the fields of language processing, adept systems, and interpretation engines. These technologies are widely rooted in various applications in different industries. For instance, in the medical field, the technologies facilitate the identification of diseases and offer suggestions for their treatments. These comprehensions are attained when data is keyed into the systems, which in turn follow up on the inquiries, memorize the facts, compare the facts to the knowledge possessed in concern to the inquiries, and offer breakdowns.

Ability to act

Artificial Intelligence systems act autonomously. Using technologies such as expert systems and inference engines, the AI systems can act or refer the actions to appropriate physical implementers. For instance, driverless cars act independently in accord with the environmental conditions since they have abilities to sense and comprehend numerous inputs. Moreover, industrial robots facilitate smooth operations in production lines since their actions are based on what they have independently sensed and comprehended.

The ability to learn

 Rather than needing encoded instructions, AI systems can gain experience and adapt their competencies. These operations are facilitated by a technology called 'machine learning'. Nowadays, artificial intelligence systems can learn rather than rigidly accept codes that previously facilitated their operations. This coded concurrently required modification in their programs in the improvement of their undertakings.

Identification of Artificial Intelligence opportunities

In terms of automation and intensification, we may categorize and describe existing Artificial intelligence solutions. Overall efficiency and performance of AI systems may be improved through automation of routine tasks. In subjects of determining qualifications, technicalities of making repetitive policy assessments may prove difficult and monotonous for humans (Kreinczes, 2016). However, these decisions are certainly easy, swift, accurate and consistent for Automated Systems.

Nonetheless, some situations can only be judged through actual human analysis. For instance, smart machines have been used in the augmentation of surgeries and fighter planes. However, it would be utterly inappropriate and impossible to automate the procedures involved in the performance of these activities. As patients would disapprove surgeries performed by 'robots', consequently automated fighter planes which operate independently and in reliance to their inhuman comprehensions, lack emotional intelligence. Therefore, the judgments taken by these Artificial Intelligence systems may not offer corporal or humane guarantees. In this accord, such fields can only require augmentation of AI systems to facilitate their mechanization.

AI Analysis Benchmarks

To maximize the outputs offered by this technological wave, businesses need to understand the criteria used to distinguish the assortment of AI solutions. Differentiations of properties comprised in the operations that need solutions, affect the preferences of AI opportunities. Whether automatable or augmentable; AI solutions need to weigh the density of the task in the operation and also, the density of the facts and evidences involved.

Consequently, the nature of the work may be predictable and direct, while the spectrum of the activities involved may be unknown and unpredictable. This consequence oversees the importance of human inclusion or augmentation of the systems involved in the actualization of the operations if the systems support. However, sometimes data may be constant, reasonably structured and in small volumes.Subsequently, the frameworks gathered in the accords of work and data complexities, four prime forms of activity prototypes are evolved. These models include; innovation, expert, efficiency, and effectiveness.

Innovation model

Artificial Intelligence technologies facilitate the identification of alternatives and optimization recommendations, while humans make resolutions and act on them. In this model, AI systems bring out creativity and human inspiration.

Expert models

The work involved in this model entails human judgment and is conditional to human familiarity and capability in the undertakings executed. Consequent to the fact that decision making and exploitation is chiefly a human duty, technology is only involved in the augmentation analysis procedures.

Efficiency model

To optimize business performance consistency, the efficiency model characterizes routine undertakings defined by instructions, processes, and rules. In the solutions involved in this model, technology discerns, understands and acts, while humans observe the precision of the solutions developed and regulate the evolution of the business environments.

Effectiveness Model

This model facilitates the improvisation of general employee and business abilities to yield the desired effects. The efficiency model informs workers on all their businesses entails. The success of this model is earned through communication and organization successively affected by aspects such as administration and sales. The actualization of these solutions may require technological interventions, where the AI systems act as assistants or agents.

Challenges of Artificial Intelligence
Human discomfort about AI Abilities

Some humans are threatened by the capabilities of AI creations. Due to their lack of emotional intelligence, their decisions may seem threatening to humanity. Therefore, businesses should be selective of their AI solutions to ensure the experiences of their employees and customers are favorable and unsuspicious

Purveyor Hype

As a result of previous unfruitful prospects of Artificial Intelligence systems, several businesses are justified to be wary of their involvement in AI systems. Developers should pitch their technologies with precision, expressing both the benefits and liabilities of the prospect. Moreover, the companies should familiarize themselves in undertakings that enhance the understanding of the solutions, before and after the acquisition.

Differences in Expansion Methodologies

Previous systems mainly based their developments on planning, analysis, designs, testing, and disposition. However, AI environments dwell on identification and purification of the research to suit the desired output. To elevate approaches in AI environments, different methodologies, abilities, and approaches need to be integrated into the environments. Moreover, intellectual systems need training in specific areas to ensure their effectiveness (Askary et al., 2018). Therefore, to attain suitable development models, AI requires different expansion methodologies such as; teaching, instructing and managing the solutions it offers.

Humans vs. technologies

As pitched by several movies, new technological ventures will create abominations that will overturn our superiority as humans, which will, in turn, lead to our enslavement by the artificially intelligent machines. However, and as previously highlighted, these solutions can exempt humans from routine and monotonous tasks, facilitating the concentration of the human workforce to more important developments. Moreover, such solutions facilitate the augmentation of other cases that require the preservation of emotional intelligence in their performance.

Societal insinuations

As new technologies evolve, various social aspects such as employment are disrupted. Therefore, different themes are experienced in the takes societies have on AI solutions. As coined by the Pew Foundation (Russell et al., 2016), societies will adjust to new ventures that only require human uniqueness for their undertakings. Moreover, the collaboration between humans and machines may certainly benefit both parties since they can learn from each other. As Google's Larry Page affirms, the internet is an example of a mutual relationship that benefits human lives and integrates new knowledge to the internet.

Artificial Intelligence develops businesses across all industries. However, the most important aspect to remember is to avoid being absorbed to the notion that any technology is the answer to everything (Banerjee et al., 2018). It is important to weigh the vitality of the works to consider the best rationale for the incorporation of technologies into complete AI solutions. Continuous discussions on the effects of these technologies on existing human activities will hasten the incorporation of AI solutions to businesses. A joint intervention of businesses, educators and executives are needed to extensively evaluate the pros and cons in AI solutions, and act accordingly. The evaluators and creators Artificial Intelligence solutions will be liable to any social effects, good or bad; will not be blamed on the emotionally ignorant technologies. However, the benefits of Artificial Intelligence are more significant to their complications.

Chapter 11:
Chatbots and autoresponders

AI powered by digital solutions to improve customer service is a new revolution, which every industry worth its salt is engaged in an attempt to transform conventional customer service in every aspect including product knowledge, brand awareness, customer acquisition, loyalty programs and after sales service.

What Are Chatbots?
A chatbot is a kind of Artificial Intelligence (AI), which uses an automated robot system that creates a virtual conversation through text chats, voice commands or both, so much like human conversations, in order to relay automated replies or carry out specific tasks for users based on a well laid down set of rules or parameters.

 Secondly, a chatbot can be used on messaging platforms, websites or applications where it can perform many functions.

Chatbot, also referred to as Talkbot or Bot in short can also be defined as an Artificial Intelligence computer program or simplified software application, designed to imitate or simulate either spoken or written human conversations or interaction with individuals, mostly through the internet in an automated manner.

It is said that the demand and usage for these messaging applications has outwitted even the most reputed networks of the social media, dating back to the year 2015. This indicates that as popularity for messaging Apps increases, more and more consumers will desire to use chats to for interaction with services and brands. By the year 2017, publicity for automated conversations increased greatly, altering brand conversations with end consumers. A good example is Facebook Messenger, which in the year 2016 broke the record of active users by a whopping 900 million.

Business customer service and sales can greatly improve through use of the now familiar messaging chatbots interface leading to heightened customer interest and consequently increased earnings and profits

How does a chatbot Work?

You might be curious, like many are, on how chatbots work. This is very important since the current trend indicates that in the near future this piece of automation system will play a key role especially so in the business arena. As a business person, you may want to try this out in planning your business as you align it with your precise business needs.

Two types of chatbot software Applications in the modern times are Artificial Intelligence Chatbots which processes natural language. This helps the chatbot application recognize and comprehend text messages or human speech, as well as discerning intent. An example is when one sends a message requesting for something, the system is "smart" enough to recognize and understand the meaning. More interesting, it even responds by asking follow-up questions such as "do you want tea or coffee?"

On the other hand, Rule-based chatbots is whereby specific commands are used in order to receive a response. This chatbot category denotes a major opportunity for enhanced customer sales service and profits for example use of some text messaging App used by some retail businesses to send offers or coupon codes advertised with a sign on a coffee shop reading "Text COFFEE to 21534 for a 5% offer."

Another example is when your phone texts "COFFEE" through the messaging system, the App recognizes the word as a command, then follows the developer rule and instantly sends back to you the coupon code. If however the word "COFFEE" is misspelt it might not pick.

Chatbots can be said to be an Application without the User Interface. To understand better let us compare an e-commerce website functions to that of a Chatbot. The former is a software APP for selling products. The layers or aspects that make it work are:-

An Application layer: A set of instructions which function-ability to the APP.

Application programming interfaces (APIs): Link the App to services so as to get for example process payments or shipping quotes.

User interface: Enables the user to inform your application what his interest is or what he wants.

Database: Stores customer data, product information, bulk of the content and transactions.

The user interface for the e-commerce website therefore plays a key role in the function-ability of the App. It has for example a cutting-edge search tool for locating products by shoppers at the same time displaying incredible product pictures. It has user buttons linked to a shopping cart where one can add products. It also has templates for entering address and payment info.

On the other hand the latter (chatbot) powered by natural language processing and Artificial Intelligence has three of the e-commerce aspects, including the application layer the database, and access to Application programming interfaces (APIs). It however does not have its own User interface but depends on a messaging platform. It can for example use Slack, Facebook Messenger, WhatsApp, or other comparable App for customer or shopper interaction. This means that the developers therefore must write links to a number of other platforms as mentioned above so as to potentially achieve a reach similar to an e-commerce website.

What are autoresponders?

An Autoresponder refers to a service that automatically enables one to send emails to different groups or a single group of people.

It is also an effective yet simple marketing tool which is used to send a succession of scheduled emails based on set specific parameters. This is manly used for follow-up purposes on ones shoppers or customers within ones store or business.

For promotion of specific items and also for new product demand tracking purposes an email can be sent to all customers or shoppers, for example as a follow-up. This can be in form of a voucher or special coupon or simply informing them of other similar products or related accessories that would perfectly rhyme with what they have bought.

How do Autoresponders work?

Autoresponders are a powerful marketing tool that is very effective and essential for internet marketers and business owners. They enable one to contact innumerable potential customers. If you consider the number of emails you are going to make just to make a single sale, having an autoresponder is a great asset for your business. Autoresponders automate up to 50% of one's marketing campaign, without which a business could lose probably a bulk of sales per annum.

Email marketing tools and Autoresponders are very important business components as follows: -

Automatic email sending

Automatic email response

Used in following up with customers.

Sending product info to clients eg. price lists and offers, newsletters

To build loyalty

To keep people in constant connection with your site.

To make sales.

Creation of email courses

Once emails are written, they are sent automatically, then subscribers sign up to receive.

Some commercial autoresponders, in addition to sending out standardized messages can send limitless follow-up messages to members, instantly without time interval delays.

Some hosting companies can provide a free autoresponder, however, purchasing an independent service creates a wider scope, to carry out more tasks like, like personalizing emails, where one can include name, phone number, address, business name, which enables attachment of a business card. Autoresponders allows creation of short email courses filled with relevant information, running for 10-15 days which can be offered to ones' visitors for free which will bring in targeted customers depending on the field of expertise selected. Article briefs are auto sent to the visitors every day. This will also create reputation for the company.

Examples of Autoresponder tools which you can research on are:

1. Office Auto Pilot
2. Campaign Monitor
3. Aweber (used for creating and sending email newsletters for marketing purposes)
4. Infusion Soft
5. Auto Response Plus (ARP Reach)
6. Mail Chimp
7. Get Response
8. iContact
9. 1ShoppingCart
10. Constant Contact

Persuasion through AI, Automation and data science
Persuasion can be defined as persuading, a term of inducement or influence. A person can be persuaded by attempting to influence their behavior, attitudes, motivations, beliefs and intentions. In the business arena on the other hand, the process of persuasion is aimed at influencing or changing the behavior or attitude of an individual or group towards an idea, object, event, or towards other persons, using spoken, visual or written words to convey information, reasoning, feelings or a combination of the three.

Persuasion is a tool that is oftentimes used in quest of personal gain such as campaigning for electoral position, in sales promotion, or advocacy. Persuasion can also be described as using ones' position or resources to change people's attitudes or behaviors.

Heuristic persuasion is whereby beliefs or attitudes are leveraged by emotion or habit.

Systematic persuasion is whereby attitudes or beliefs are leveraged by reason and logic.

Automation is a technological procedure or process performed with nominal human assistance. Also referred to as automatic control uses a variety of control systems in equipment operation. Equipment like machinery, factory processes, ovens for heat treating, telephone networks boilers, stabilization and steering of ships and aircraft, vehicles with reduced or minimal human intervention and other applications.

Automation includes applications ranging from large industrial control systems to household boiler controlled by a thermostat, from high-level multi-variable algorithms to simple on-off controls.

Automation has been applied mostly in combination in mechanical, pneumatic, hydraulic, electrical, electronics and computers. Complex systems, like airplanes, ships and modern factories, and typically apply these combined techniques. Automation benefits include savings on labor costs, savings on material costs, savings on electricity costs, and improvements to accuracy, quality, and precision.

Scientists help in retrieval of valuable information from a sea of data to be examined and the discoveries used to streamline the companies. The role of data scientists is to analyze data, asking information-driven questions, applying mathematics and statistics in order to discover important outcomes.

In our current digital world where there is a remarkable increase in automation, there is an urgent need for us to initiate conversations and dialogue, especially so in recent times where we are confronted with the amazing field of Artificial intelligence. What with smart voice chatbots, autoresponders, Netflix, self-driving cars and the field of robotics, just to mention but a few. For us to embrace Artificial Intelligence (AI) for example, we have to be made to understand not only how the different components work, but how it affects society. Oftentimes, persuasions regarding automation are mostly centred on the job market, its negative or positive impact, more so as AI systems become increasingly commonplace.

Two ways in which the current trends in automation will influence and affect our future is that for one, it will in no doubt lead to a more efficient and superior future as witnessed by industrial revolution- jobs will be lost yes but due to rapid expansion, new openings will occur for better more advanced and efficient jobs.

 The other way is to recognize that the present times are extraordinary with robots becoming increasingly fit and smart. It is possible that the number of job opportunities and enterprises they will consume or crush will outweigh by far the number of job openings they will create.

Whichever of the two possibilities above turn out to a reality, what we know for sure is that automation or AI for that matter is drastically influencing the economic landscape in these extraordinary times. We are in realization that a big percentage or portion of our lives and jobs are increasingly being automated.

This concern is not new for during the nineteenth century, textile workers in England are in record to having destroyed weaving looms, these workers were then accused of being opposed to technological advancement.

The talk is rife of the possibility of intelligent machines in future turning against humans, or their makers for that matter. Potential for disruption of the labour markets by intelligent machines has been examined through academic studies and various books published most recently. Furthermore, detailed studies have been carried out on a possible serious threat posed by automation to human existence altogether, with special reference to AI.

The AI technology could not have been predicted better than through a classic science fiction Space Odyssey film of 1968, directed and produced by Stanley Kubrick. The film, following a space voyage to Jupiter discovers beneath the Lunar surface a mysterious featureless artifact or alien monolith. Mankind then sets off on an expedition to discover his origin using HAL 9000, an intelligent sentient (perceptive and intelligent) supercomputer.

Automation is nothing new of course for robots have been in use for decades in industrial assembly lines. The current automation wave draws its benefits from affordable computing power with emerging new areas in software application such as image and language processing. This new generation automation machines and AI systems may adversely affect white collar jobs which was not previously so. Apart from potential consequences to employment, there are other widespread implications in other areas such as education, privacy, cybersecurity, healthcare, environmental management and energy.

Cheap bandwidth for example is influencing learning methodologies which can be seen through online platforms which is greatly influencing the how and what is learnt. A good example is MOOCs (Massive Open Online Courses). As automation increasingly takes a big chunk of routine tasks, this kind of learning which stimulates creative and conceptual capacities becomes more relevant and popular, consequently leading to a shift in the education system from conventional reading and mathematics to more focus on intellectual and personal skills, working hand in hand with intelligent machines akin to AI.

Universal individual and commercial data collection and storage from social media platforms on the other hand raises privacy concerns. Yet another area of risk is Cyber-security. Developed economies are feeling the impact of automatization more than developing countries which are following closely behind especially so through investors.

Customer service: reduce workloads and handling retention, segmentation and scoring

Two benefits of application of AI in customer service is workloads reduction and customer retention

There is increasing demand globally for excellent customer service due to exposure to more advanced products and services. Customers' expectations are rising demand for seamless service is increasing day by day, especially so with more and more automated services that are technology aided, social platforms, messaging services and applications.

Customer retention is key as it boosts revenue, thus businesses target is to build loyalty and in so doing retain customers. It goes without saying that good customer service leads to customer satisfaction which translates them into brand ambassadors. They carry out free brand promotion to family and friends, consequently boosting sales which increases revenue.

It is proven beyond any reasonable doubt that poor service sends away customers and on the flipside they return if they receive good service consistently. Customers are prone to adept and efficient systems and cannot prefer an inefficient, outdated system when there are alternative advanced channels which are faster and time saving akin to automated systems such as AI can offer.

Artificial Intelligence (AI) therefore combines intelligent services with automation in order to achieve accuracy, scalability and efficiency which manpower alone cannot achieve.

AI can be tuned, trained and customised to fit business processes in the current times. AI-enabled customer service from contact centres improve or enhance human agent capabilities supported by AI chatbots provide customers with a wholesome service experience. AI chatbots can handle common customer queries and even suggest solutions while human agents can handle more complex ones.

Predicting Consumer Behaviour and Referral programs

With emergence of Artificial Intelligence (AI) could help simplify understanding of consumer needs and wants ideally even before they even do themselves. Deep learning, being a sub-set of AI can potentially transform marketing in supporting prediction of consumer behavior by businesses. This machine method of learning uses deep or layered neural networks, comparable to biological brains able to not only acquire or learn skills but also in solving complex problems at a higher rate than humans. It helps robots or computers to perform "human" tasks like recognizing voices, translating languages and perceiving objects. With a set of inputs, deep learning helps train AI in predicting outputs. However deep learning requires large data and computational power, though much less data pre-processing by humans is required as compared to machine learning techniques. With access to key elements however, a deep learning system can learn human behavior prediction pretty accurately.

Referral programs are becoming a widespread way of acquiring customers. However there isn't much to prove that these nature of customers carry more weight than regular customers. The question still lingers whether the referred customers are more loyal and more profitable and if so, to what extent? Researchers have attempted to determine this.

A leading German bank was tracked approximately 10,000 customers for a period of about three years. The findings were that referred customers:

1). Show a higher contribution margin rate which however eroded slowly by slowly

2). Indicate a higher rate of retention that the regular customers which persists over time

3). comparatively in the short and long run they are more valuable than the counterpart customer of similar time of acquisition and demographics.

A referred customer recorded higher average value by 16% than the non-referred customer

Up-sell and Cross-Sell with AI-Powered Product;

Upselling refers to a product promotion using a complementary option for example after one buys a computer the customer is given an offer of alternative memory options or processors ("we have a higher caliber computer than your selection, would you consider adding some little cash you have it at a discount...."), while on the other hand in Cross-selling the customer is encouraged to purchase a different product with their original purchase, say accessories ("can you also take this flash disc at a discounted rate which is very crucial...?). In cross-selling, the complementary products do not have to be those that work together with the purchased product one can suggest related products not necessarily those that work together. Sales services too such as priority service or warranty is also a cross-selling strategy

Use Predictive Analytics to Provide New -Product Recommendations and Boost Revenue, data-driven approach

Predictive analytics is a combination of predictive modelling, advanced analytics, real-time scoring, data mining and machine learning which helps identification of data patterns by companies. Historical data is used to predict future trends in business for example. This is achieved by feeding the historical data into a model which analyses the data and identifies patterns or trends. The model studies the historical data, which is then used to predict or forecast future outcomes by applying it to current data.

Predictive analytics is currently used widely in many ways such as: -

Product recommendation system which predicts customers' likes

Customer Lifetime Value (CLV) measures predict the estimated amounts a customer will purchase form a company

Others are; Predictive maintenance; Optimizing marketing campaigns, credit scores, fraud detection and sales forecasts

According to an Accenture survey conducted recently, it shows that since 2012, most company executives have placed predictive analytics high on their executive agenda. This is due to increasing awareness on predictive analytics indicative of potentially profitable future outcomes.

How do predictive analytics work?

With large amounts of data, the way relevant insights are derived from information has greatly changed. The approaches previously used referred to as Business Intelligence (BI) which used typical structured data sorting technics have been switched to techniques which use raw information. While the traditional BI uses deductive approach where the assumption is that there is some relationship or understanding between existing patterns with relationships, the Inductive approach on the other hand is more concerned with data thus doesn't make any presumptions about patterns, relationships and discovery.

Predictive analytics thus use the latter- inductive reasoning applying it to big data through machine learning, Artificial Intelligence and neural networks, to identify patterns and interrelationships.

Predictive analytics can be used in: -

Business problem solving

Decision-making

Optimizing processes

Cutting down operational costs

Improving customer experience

Market opportunities identification

Risk reduction and mitigating through problem prediction.

In a company's sales department for example, staff ordinarily rely on their expertise, knowledge, professional experiences and even intuition or "gut feeling" in decision making. For example, they know their regular customers off head, customer's interest or preferences, which products move and so on.

Some or most of this knowledge is based on guess work through intuitive choices or even personal preferences.

If such a team therefore is enhanced with AI, actual data is used to help them come up with more accurate and efficient predictions which in some cases confirms their previous intuitive choices to be true. In many cases they will watch with awe on huge quantities of data unlocked by automation or AI for that matter which would have ideally been omitted.

Another advantage is that in the event where an expert leaves, which is usually the case, the AI model stays together with the unlocked knowledge so no need to panic.

CHAPTER 12: THE RIGHT ARTIFICIAL INTELLIGENCE APPLICATIONS FOR YOUR BUSINESS

It's very important to understand how Artificial Intelligence can be of positive impact in your business. There are many AI applications that can be used in businesses to make running of the business much easier. These are several Ai applications that when suitably applied can play a key role in reshaping of your business.

 Automated systems that are installed with AI can greatly help you to make better decisions in managing business resources. There is a wide array of Artificial Intelligence applications that can be employed in business for beneficial purposes. Some of the Artificial intelligence applications are used in different industries including; healthcare, customer support, finance, security, drones, smart cars, creative arts, among others.

In the present day, most of the customer questions or complaints are responded to by human beings; this is done via telephone calls, emails, social media dialogues and online chats. However, AI applications have been absorbed in business systems which has facilitate the automation of these communication. Computers may be installed with Artificial Intelligence applications to enable them to respond accurately to customers. Further, combining AI and ML enables the AI machines and computers to work more efficiently. Choosing whether or not to use AI in your business will be determined by what you hope to achieve, in the long run.

How are artificial intelligence applications applied in accounting in business?

Business accounting refers to the systematic recording, analyzation, interpretation and presentation of financial information. In small capacity businesses, accounting is usually handled by one person, or by various teams in larger organizations.

The accounting process is of great importance in any business in that it helps the business owner to keep track of the business' operations. Accounting process helps in analysing finances which enables the owner to make wiser decisions, especially in handling finances. Incorporating Artificial Intelligence applications in businesses makes the accounting work easier, which enables the business owners to honour their compliance obligations.

Artificial Intelligence applications are set to transform the accounting and finance industries; there is the elimination of tedious tasks which translates to saving time. Employees can therefore engage in other tasks that have more productive impact on the business.

HANA is an Artificial Intelligence application that helps businesses to convert their database into intelligence tools that are more useful and practical. The application works by ingesting and replicating sales transaction data drawn from relational applications and databases.

In drawing deeper insights and streamlining finance processes, firms need to explore current AI applications. These applications assist accounting managers to stay ahead of business transactions in the midst of time-consuming and tedious processes. This is made possible by making use of the Machine Learning app instead of spreadsheets and PDFs. Receipt images are extracted and automatically classified by the Machine Learning app; the extracted data is classified bases on the spend category. These reports offer smart insights for businesses which help in better financial planning.

The ML app is a branch of AI and it yields deeper insights in that data processing is done over time; this provides businesses with comprehensive overviews on the spending patterns which can be used in making better finance decisions for companies. The basic AI apps are undoubtedly key players in the business accounting field.

Are Artificial Intelligence Applications used in founding for startups and established companies?

Startups have been dominated by various software giants including SAP, Oracles, IBM and Salesforce. They provide various applications that are applied in customer relations management, resource planning and human resource management. Several startups are offering the next group of innovative services via Artificial Intelligence apps,

Companies that are powered by Artificial Intelligence, many of which are AI startups are making use of these in their recruitment processes. For example, Salesforce has made investments in DigitalGenius, which is a management solution for customers and in Unabel which offers translation services for businesses.

Artificial Intelligence startups are now generating solutions for various industries. These can be seen across several business sectors i.e. legal, automotive, agriculture, healthcare and finance. It's very clear

The Artificial Intelligence startups are becoming very popular because of their capability to provide established companies with prized point solutions; this is mainly due to the fact that they can access domain knowledge and large data volumes.

A majority of AI startups are developing some type of Machine Learning models that have the ability to predict or classify end results based on certain input data. In majority of AI startups, there effective running is highly dependent on data volumes; the larger the data volumes, the better their performance.

Artificial Intelligence in Creating the right team to lead success in business

The rapid increase of Artificial intelligence in business has been brought about by the need to answer various questions in the recruitment process i.e. the exact skills you are looking for as you are recruiting, the organizational structures that will work best for your business, the organization module that will result in success while using Artificial Intelligence, etc.

Many businesses have realized the importance of implementing Artificial Intelligence teams that play important roles in improving productivity and service delivery to customers.

Making use of AI teams that are well-rounded enables businesses to reap maximum AI benefits. In the process of selecting members for the Artificial Intelligence team, it's important to consider reorganizing your recruitment processes.

You will require talent that can grasp computer learning processes. A centralized team is basically what will be the driving force behind the Artificial Intelligence applications. Always remember that the main aim of Artificial Intelligence is to basically reduce work in the shortest time possible. Having the right team in place will facilitate better performance of the business which is bound to reflective positively in the long run.

The effect that having the right team in place in any given organisation can never be underestimated. This is why employers need to employ the useful AI apps to enable them to recruit people with the right skill which ensures that tasks are performed within shorter time frames which can greatly help in meeting targets and improving productivity.

The Artificial Intelligence team is tasked with executing the whole AI project successfully; this means that the team must have engineering talent for them to carry out this task successfully. Depending on the project that needs to be implemented, data engineers, data scientists, machine-learning engineers and product managers could be among the AI team members.

Baidu and Google are excellent technology teams; they have proven that they have the capability to function cross-functionally in creating strong AI value via advertising, product recommendations, speech recognition, etc.

How are Artificial Intelligence applications used in the collection, preparation and recycling of data?

Data is generated in huge volumes in the business world. These can be either free text, categorical or numeric data. Data collection can be defined as the procedure of gathering information and measuring it from several different sources. If we are to utilize the collected data in problem solving in our businesses, there must be sensible collection and storage of the data. These will facilitate the development of practical Machine Learning and Artificial Intelligence solutions.

The procedure of data collection and preparation is very vital in business since it enables you to have records of past events; the data analysis then kicks in when looking for recurring patterns. Predictive models are then built from these particular patterns which can be used in predicting possible changes in future.

Good practices need to be observed when collecting data to facilitate the development of high-performing models; this is highly dependent on the data that is collected. The collected data must be free from error and contain pertinent information to facilitate the performance of the set task. The collected data is then re-used for laying strategies for solving similar problems or tasks in future.

Artificial Intelligence analytics solutions for companies

Artificial Intelligence analytics solutions possess the means through which value is derived from the rich information generated from the collected data. Progressive analytics platforms enable businesses to create insights on business processes as well as predicting outcomes; this is based purely on information that is collected from data in the past. Artificial Intelligence on the other hand incorporates Machine Learning capabilities by acting as the force multiplier for the generated insights.

Artificial Intelligence applications and analytics solutions aim at making businesses more productive, effective and efficient by providing valuable enterprise insights in an accessible manner. These insights are linked to the most important goals of the enterprise.

The best analytics software and AI provides leverage for ML algorithms in data platforms which transforms big content and big data into data visualizations; these help the business to increase operational efficiencies, maximize revenue and increase automation.

Open Text Artificial Intelligence and Analytics provide a detailed set of analytics tools and AI which help businesses to sort out data challenges at a comfortable pace. This warrants a clear path towards Artificial Intelligence technologies whose main focus is on improving decision making, business optimization and automation. This can only be achievable with the appropriate tools for generation of insights. The main achievement is the identification of trends, patterns and relationships through analysis, processing and recycling of data. Businesses become better equipped to handle similar situations in future which leads to improved productivity for the enterprise.

Some of the positive impacts of Artificial Intelligence and analytics platforms include:

- Increased operational efficiency
- Cost reduction
- Increased productivity
- Accelerating the rate at which AI value is delivered in a cost-effective, time-saving manner
- Improving decision making and visibility
- Maintaining operational improvements continuously

Business strategy in the Artificial Intelligence economy explained

Breakthroughs in the business economy related to the artificial intelligence program has led to the redefining of business operations across the world. Business strategy and Artificial Intelligence initiative has encouraged the growth in the use of AI in the enterprise landscape. Specifically, this exploration looks at the manner in which AI affects the execution and development of business strategy.

This particular initiative is aimed at reporting and researching on the manner in which Artificial Intelligence has resulted in a drastic workforce change, privacy, data management, the generation of new business challenges, and cross-entity collaboration. It investigates new threats and risks in security, job loss and dependency as well as seeking to assist the business managers to understand the huge opportunities from combining machine and human intelligence.

An example of the manner in which business operations have been refined is the manner in which data is collected, processed and used in the optimization of customer interactions. The alteration is playing a key role in the redefinition of structural dynamics of the management concepts, economic theory, and business strategy. Leading companies in the technology field globally have cast their nets wider in their research with regard to Artificial Intelligence.

AI has the ability to help in transforming businesses just in the same manner that the internet has transformed the way businesses is done. From smarter services and products to automated and optimised business processes and better enterprise decisions, Artificial Intelligence is well able transform everything in the business world. Businesses that aren't capitalizing on the Artificial Intelligence transformative power risk being excluded in the growth process.

In conclusion, there are several Artificial Intelligence applications that are in use in businesses. The main aim of implementing is the businesses is for the possible benefits associated with me. It's critical to do a proper analysis of your business and the Artificial Intelligence applications to determine which apps will work for your business and those that will not.

Cost implication is the major concern; many businesses when at the implementation stage of Artificial Intelligence lack adequate personnel. This mean that they have to outsource for manpower to ensure that the system delivers the intended results. Another setback is the cost implication in regards with the maintenance and implementation of the AI system. Sourcing for AI applications that are pocket friendly is a great idea in cost reduction. Take time and find out the applications that are being used in businesses that are similar to your business.

CHAPTER 13:
ARTIFICIAL INTELLIGENCE FOR MARKETING AND ADVERTISING

Artificial Intelligence in Marketing and advertising: a revolution for your business

Artificial intelligence marketing is a technique of grasping client data and AI models like machine learning to predict your client's subsequent action and to improve the client 's experience.

Key components of AI Marketing

AI based marketing is a very powerful tool today that we cannot ignore. It comprises a couple of aspects as listed below.

Big Data

It refers to a marketer's ability to combine and fragment huge groups of information with least labor-intensive effort.

Machine Learning

They are crucial when analyzing large volumes of information. This is by pointing out patterns or repetitive occurrences. Also they help in accurately foretelling ordinary acumen, feedback, and opinions. This in turn helps marketers comprehend the origin and probability of certain undertakings recurring.

Powerful Solution

This means that they identify intuitive ideas and themes within large data groups, extremely fast. Also they deduce feelings and communication like a person, which enables them to comprehend information from social media, and email responses.

Digital optimization and personalization

Digital optimization is the action of using arithmetical technological expertise to advance current working routines and corporate designs.

The progression of hugerecords together with progressive logical results have made it promising for venders to form a vibrant image of their goal audiences than in the past; and in this hotbed of advancement lies artificial intelligence (AI) marketing.

Equipped with large data intuitions, arithmetical salespersons can significantly improve their crusades' results and the return on investment. This can be realized with fundamentally no further work on the Salesperson.

What marketers do better than machines

Once it comes to numerical breakdown and quantity chomping, technology is the best option if you're operating *with* it, but then again your worst rival if you're contending against it. Technology is faster and more precise than people, and its inferences cannot be influenced by individual or emotional preference.

The theoretical inference is that we ought to turn most of our choices, likelihoods, diagnoses, and verdicts, both the inconsequential and the consequential, over to the procedures. There's is no argument at all about whether doing so will provide us with superior outcomes.

Why people will keep on being essential to marketing

We are not existing in those revolutionary sci-fi cinemas so far! Apparatuses are compliant, we authorize them whatever we want them to do, and don't have a brain of their own, so they definitely don't have the expressive intellect to attach with a human.

Machines can't distinguish what's "hot" and anything's that is cold. Movie marketing is a picture-perfect illustration of this; a device remains incapable to perfectly guess which movie character a precise individual will hunger published on their outfit, based on a trouser they bought in the previous summer. Human specialists use a combination of their personal emotional intellect and analytics to guess this. Some people's likes and favorites are puzzlingly qualitative and unplanned. They are frequently not demarcated by relative groupings, but reasonably by random and unscientific aspects, the summation of a lifespan of irreproducible involvements and the memories that linger. This marks such a guess unachievable by a machine.

Machines do not have a soul

How does one recognize when someone has taken their time to do something? You don't even know how, but one can make out even the well-crafted computerized email. How do you tell? It's probably be the lyrics used, the sentence structure, or even the layout, but one way or another you just discern it's a machine. In some way, we comprehend the emotional limitations of technology, and this makes us important to communication with our audience

In what way is marketing technology intended to support humans?

Technology isn't intended to substitute human, it's intended to help you as a marketer. When you recognize how to leverage technology, you will join the greatest, fruitful and profitable products.

As a matter of fact, advertising technology is undertaking the hardest part of your marketing progressions, like segmenting your clients and diagnoses and directing them with content that is contextually applicable to them at a certain time. This gives you the chance to turn your devotion back to the skills of influence, and to get involved with your clients on an expressive level that apparatus basically can't.

Here are several vital takeaways one ought to know about where this expertise can go and in what way it can benefit you.

Machine knowledge will at no time substitute humans as the leading spring of inventiveness; it will simply assist make that content manufacture further effective.

The total amount of content and information to be created in the subsequent number of years will exceed the human brain capability to create and compute.

Machine knowledge will offer products with the supremacy to put their Huge Information to use and control their analysis rather than allowing the data slip up through their fingers.

Machine knowledge will aid in creating a more modified involvement for the operator, refining his or her impression of the product.

It will assist brands create their marketing hard work more effective by enhanced targeting clients.

In spite of that, there is still a crucial requirement for human inventiveness and awareness, meaning there isn't an abrupt danger of digital dealers becoming outdated. Yet, when approaching the overview of fresh technologies, it's essential not to oversee administrative fundamentals and the human level in edict to ensure operational implementation and operative satisfaction. Those firms that have moved fast to revolutionize the whole client expedition by leveraging modern t technological advances, whereas also maintaining the human level in mind, will be the ones to win the contest for digital mystery.

The correlation amid humans and machines has transformed drastically in latest years, and will keep changing swiftly in coming years. It's at a critical moment at which technology has been extensively accepted as the enabler of commercial progression and fast-paced invention, letting us to gage processes and manage high capacities of information.

Marketing platforms and tools for your company.

Before choosing AI based marketing platform there are a couple of factors that you have to look out for. They are as follows

How user friendly is it?

You need a marketing platform that is easy to use and navigate. This is because as a marketer you need to spend as minimal time as possible maneuvering a said platform. It saves time and is easy to learn and master. Also consider the users and how well they interact on the platform in regards to timeliness delivery of reaction, response and feedback.

Is it compatible with other marketing platforms that you have?

Before you purchase any AI marketing product you need to consider other products that you already have. How compatible is the new product with the pre-existing ones? This ranges from the content management systems (CMS) to any email marketing software.

As a company, is there enough data to research from?

The efficiency of any AI platform is contingent on the volume of data that is fed to it. The more data there is to research from the more accurately will it address your problem.

How well does it address your particular problem?

Even before sourcing for an AI product are u certain that you have narrowed down your specific problems and if so how accurately does this product cover or solve this problem

Who are the developers of the product?

In these times of inventions and new discoveries there are a ton of developers out there. However, when it comes to settling for an AI product, you need to research on the developers and be sure that they are people who are conversant with the market and its' evolving trends.

Having the above in my mind, below is a couple of marketing platforms and tools that may be suitable for your company. We will divide them into five entities that are based on the purpose and area of use of each.

Artificial intelligence product for content generation

Content generation process is key to marketing. It is not an easy task and marketers can attest to it but not to worry since AI products are now available. They come in handy by scrutinizing existing content and find data based methods of making it better. Some of the products under this category are;

Acrolinx

Acrolinx assist you in making certain that the content generated matches with your strategy. You feed the tool with any instruction that you have generated, and it gives real-time feedback. This feedback is to confirm whether your content matches the instruction. Eventually, the product researches from the content it scrutinizes and is able to give better ways for how to better the content you create.

Brightedge

Brightedge is a Search Engine Optimization that has an AI fueled attribute called Insights. With this tool you get better results. This is because of its ability to go through tons of web pages providing precise action items to clients they to improve their SEO, including suggestions on methods to update content creation.

Artificial intelligence product for content strategy

Content strategy development is key to good content marketing efforts. This is done with a combination of human intuitions and data that is effective. By using AI platforms marketers can now analyze data to generate more precise facades, realistic goals, and more formidable content plans.

MarketMuse

It is based on learning and natural language processing. This is by scrutinizing a corporation's data comparing it to multiple websites based on the same content. From the analysis it gives suggestions for content improvement as it is generated.

Concured

The product uses data and machine learning to recommend content topics founded on audiences reacts to responds to, finds loop holes in already available content in the market and give guidelines what content to promote.

AI Products for Personalization

Individuals progressively anticipate for a tailored marketing experience . But there are some AI products predominantly fixated on using machine learning to aid salespersons convey experiences to buyers tailored to their interests.

Drift

It offers AI-driven chatbots—an altered way to generate a tailored experience for guests to your website. Using natural language processing machine learning chatbot can decide the most appropriate responses to the enquiries guests have in actual time.

Uberflip

It uses AI to decide which of the content fragments that will be the most suitable to each guest centered on data about their actions and interests. The tool programs the procedure of generating an appropriate experience for respective individuals that frequent your website.

AI Products for Email Marketing

This is one of the most vital virtual marketing strategy. This is due to its remarkable ROI (return on investment).

Optimail

It is an AI product that helps you describe your key email marketing goals.

Phrasee

It is an AI product specifically dedicated on writing better subject lines.

AI Products for Promotion

In any website you need as much traffic as possible. With AI tools can now help you get improved outcomes for your promotion campaigns.

Cortex

It's an AI product that evaluates loads of data topics from different companies studying what the public react to. From this information, organizations can now know when and what to share on social media to achieve more impact.

Albert

It's a self-learning marketing platform that programs the conception of promotion campaigns centered on previous data.

Acquisio

It's an AI tool for handling PPC campaigns. By using AI, the campaign creation process is simplified maximizing campaigns for better ROI.

PERSONALISATION OF CUSTOMER EXPERIENCES

The capability to generate appealing connections through digital networks with clients has a major effect on progress, returns and product activism. It is accurate to note that feelings can be applied to monitor corporation digital scheme for building digital network associations with clients.

AI is aiding convey improved consumer involvements by creating improved merchandise recommendations, aiding quicker sales and enabling more appropriate personal shopping.

ANITICIPATING CLIENT PROSPECTS.

What better way of conveying improved involvements than by forestalling clients' hopes and provide exactly their desires in a timely manner.

Various businesses use merchandise commendation engine to find most consumed client favorites and merchandise bring them to a brick and mortar setting.

CUSTOM-MADE COMMENDATIONS.

Clients who are tech-conscious customer require personalization and vendors are rising to the task AI, machine learning and large data.

AI can suggest products based on recurrently purchased things or similar products.

CONVERSATIONAL ADVERTISING

Conversational marketing is basically a feedback focused on marketing tactic embraced by enterprises to motivate client base and client reliability.

For a business to mature and continue being important, they require to establish a smooth link with the customer, influence the objective audience, and increase client conversations.AI is the apparatus that controls conversational marketing

The client of today doesn't want to tend unfriendly calls, personal communication; they want swift and custom-made information. Voice facilitated conversational advertising because it can give direct responses and feedback

By using conversational marketing it can assist condense the scales sequence by providing instant and important responses to probable customers and reducing on the time required on taking and succeeding leads.

Conversational marketing is altering the way business dialogue with their clients. It overlays the way for collaborative and modified communications .AI powered conversational marketing scrutinizes through the selling noise and provides important communication at the precise time.

SALES AND MARKETING FORCAST

Some of the major benefits of AI in digital marketing is its capability to foretell client activities and predict sales. Using prognostic analytics, AI puts together data mining, numbers and auction modelling to foretell future results for online businesses.

AI aids teams to close extra business with bigger distinguish ability into their determination and effectiveness. It gathers auction activity data to understand where auctions teams use their time then use that data to find stagnant contracts.AI provides timely warning signals about pacts that are slowing down then constructs an activity founded achievement roadmap for every single chance.

AI uses this entrée to data to provide AI compelled sales coaching to aid auction reps decrypt why contracts are wedged and what to do subsequently qualifies them to target themselves beside top performers, and recognize what brands top performers unlimited.

AI has established a comprehensive behavioral analytics result that defines which auction rep activities that are most probable to close a contract.

1. CRYSTAL IDENTIFIES

Crystal provides sales experts character profiles for everybody they come interact with. It plans profile material across the web, in gears like linked, sales strength and more. Crystal grants them access to character motivated email template centered on the recipient's character. This sort of custom-made communication is intended to resonate with receivers, extending the relationship amid the auction pro and client or prospect.

Crystal works in cooperation with web and desktop founded software solicitations, where it actually shines in email submissions.

Crystal can bid suggestions on the dialect to use as auctions pros type, suggests phrasing ideas and template directly in the tools, making it easy to use

Crystals AI gives character perceptions at the point of necessity, facilitating sales experts mend their communication with client's prospects and colleagues

2. TROOPS

Allocated as slack bot for auctions force, troops aid sales experts mechanize their work flow and line them up with well-designed groups like marketing and merchandise management. It assists to pull info from auctions force and makes it stress-free to publicize, and changing a slack channel.

From auto report generations that are available in slack to looking for sale force client info in a specific slack frequency, troops merge data for sales experts in an individual spot. Sales supervisors like troops for cool, no- coding necessary way they assimilate sales force in slack, forming new information, control panel and alarms that keep their crews up-to-date on what is going on in actual time.

Troop's aid sales experts uphold their sales duct more effortlessly as it spontaneously tracks sales force information and bestowing it to them in a laid-back manner. Several clicks and it organize the rest.

3. QUILL

From narrative science Quill is a regular language producer for creative organizations including sales crews. It evaluates online and digital information to categorize the facts, lyrics, and linguistics that are vital to your auctions organization. It yields content of your auctions organization that meets your corporate guidelines and style inclinations like character, style, structuring and the lyrics you use.

Quill intensifies the importance of the sales data establishments   have already and generates modified sales writings, reports etc. That liberates your sales team's time.

4. CLARA LABS

Clara is computerized and spontaneous meeting scheduler. She uses dialect to reply to emails demands, is accessible twenty four hours in a week, and monitors up spontaneously with meeting parties.

Clara saves sales experts time and work, reserving meetings simply and without difficulty.

MARKETING FUNNELS AND LEAD GENERATIONS

AI is transforming lead generation and dialog. In the long run, there will be a great transformative impact on businesses and vocations.

AI structures are hugely centered on colossal data volumes analysis. Once they analyze data they now apply those insights to generate predictions. Also AI systems learn and progress over periods on their own as they are exposed to more data.

We could say that AI integration in marketing funnels and lead generations create platforms for organizational and careers growth at an impressive rate.

ACQUIRE VITAL INSIGHTS ABOUT LEADS

Huge data sets can now be analyzed in an instance thanks to AI technology and thus more accurate decisions and predictions can be reached at.

AI can analyze previous lead patterns than convey to you which leads to pay attention to.

BUILD YOUR LEAD DATABASE
Based on information about current leads, AI tools can find you more leads hence emphasizing on the importance of AI integration.

AI SOCIAL MEDIA ENGAGEMENT
Compared to various facets of people's lives, AI manipulates online media advertising in various dimensions. As a result there has been huge impact of Artificial Intelligence in the segment. Also a number of the top major leading major labels have joined onboard.

Because of cumulative request for public grid advertising which utilize smart systems. Numerous companies have upped their game to offer this provision. Currently several corporations are aiding business to grow by involving objective viewers in public media podiums.

APPLICATION OF PICTURE RECOGNITION IN ADVERTISING
Facing escalation of machine intelligence in public mass media marketing, pictures have a different resolve. Intelligent tech has aided public network dealers enabling efficient image usage to increase commitment rates. Additionally salespersons now have picture recognition software to find forms in customer activities. By utilizing AI enabled software merchandise and amenities agents aim at definite clients modifying content on their portals.

INTRODUCTION OF FACE IDENTIFICATION IN SOCIAL MEDIA MARKETING

Through machine intelligence public media salespersons can now incorporate facial recognition in public media marketing. Example of Facebook it uses face recognition to make the tagging purpose easy. Facebook recognizes the individual on the picture spontaneously saving the user the time to type the name of the person on the image.

Salespersons can benefit themselves from this feature by spontaneously tagging persons on the images of their merchandise, badges and so on.

UPRISING OF CHATBOTS AND VIRTUAL ASSISTANTS

Chatbots have aided marketers on public sites in various manners. This is by removing all presumptions from promotion. Formerly promoters had to chance when pushing their merchandise on public forums. This is because of inadequate information on what was efficient or not.

They have aided promoters to conduct in-depth scrutiny of different postings on public networks that are in line with their scope.

Chats have made it possible for promoters to forecast the likelihood of their posts doing well on various public forums.

IMPROVEMENT IN THE SOCIAL LISTENING PROCESS

Social listening can be defined as the procedure of tracing dialogues about certain words, slogans or trademarks and from the data collected to discover openings, compose information aiming a certain client segment on public media forums.

Using public location listening, salesperson can discover operational methods of connecting with clients in actual period and consequently study ways of molding promotion crusades to reach the customer more efficiently.

DIGITAL MARKETING TECHNOLOGIES

AI tools can aid generate content schemes that are resourceful supported by data. These schemes are

Hubspot

Hubspot utilizes machine learning to aid salespersons realize fresh content concepts that thrive and certify those content concepts in best proficient methods available.

Concured

It shows salespersons precisely what matters motivate commitment and what to scribble subsequently. The outcome is AI driven content scheme forums that mechanizes content review, subject search the generation of data based content guidelines content advertising and performance evaluation.

Crayon

Through machine learning to offer viable acumen on precisely what your rivals are applying online. You can monitor how the key folios of a corporation's website transform gradually.

Acronx

Acronx is a content approach placement forum that aid main enterprise trademarks better the value of their content at measure. By applying AI to go over content and ensure it suits your requirements regardless who the composer is.

Vennli

It based on content acumen forums that help marketers to be more appropriate to consumers in messaging, content strategy and marketing infrastructures.

Vennli grants the marketer the ability to gauge their content logistics over time, in order to measure performance at scale.

MARKETING FUNNELS AND LEAD GENERATIONS

AI is transforming lead generation and dialog. In the long run, there will be a great transformative impact on businesses and vocations.

AI structures are hugely centered on colossal data volumes analysis. Once they analyze data they now apply those insights to generate predictions. Also AI systems learn and progress over periods on their own as they are exposed to more data.

We could say that AI integration in marketing funnels and lead generations create platforms for organizational and careers growth at an impressive rate.

ACQUIRE VITAL INSIGHTS ABOUT LEADS

Huge data sets can now be analyzed in an instance thanks to AI technology and thus more accurate decisions and predictions can be reached at.

AI can analyze previous lead patterns then convey to you which leads to pay attention to.

BUILD YOUR LEAD DATABASE

Based on information about current leads, AI tools can find you more leads hence emphasizing on the importance of AI integration.

20 Tips to Use Artificial Intelligence immediately for Your Business

The trick in the current success in business today can be linked to the incorporation of Artificial Intelligence systems in the day to day running of the organization. Some of the tips in which you can make use of AI for business growth include:

- Use Analytics that are Artificial Intelligence based to enable you to make better decisions for your business
- Add automation in your structure to increase your marketing strategy which translates to increased sales
- Make use of supply chain management to help in better inventory management
- Use AI to improve the safety and maintenance of your business equipment
- Use AI systems in your hiring process; this will enable you to recruit the most qualified candidates for your organization
- Use AI to protect your business from cybercrimes and in fighting fraud; if not checked, these can lead to serious negative business impacts
- Incorporate AI systems in your business to help in expansion via self-driving technologies
- Use AI for customer support optimization using a chatbot
- Use AI in object, facial and image recognition
- Use AI to create smarter end products
- Do your research to find out the appropriate AI apps that will give you maximum benefits
- Use AI to deliver customer experiences that are personalized
- Use social media platforms that are AI powered as your marketing tool
- Use AI to increase manufacturing and output efficiency
- Use AI in optimizing business logistics
- Use AI in predicting performance
- Use AI in predicting behaviour
- Use AI in advertising and marketing your business
- Use AI in analyzing and managing your data
- Use AI in automating workloads

It is accurate to say that integration of Artificial Intelligence in marketing has greatly revolutionized marketing. We can only speculate and be optimistic on what the future holds for AI based marketing as the impact is already felt and we know that it can only get better as time and technology advances.

CONCLUSION

Artificial Intelligence is one major component in business in the modern world. Current trends indicate that AI is going to be even more incorporated in businesses in future. Important to note is that for the effective incorporation of AI into your businesses' analytics program, it's vital that strategies surrounding it goes hand in hand with the overall strategy of your business, always considering the merging of technology, people and processes. Artificial Intelligence is helping to input "increased smartness" to machines but it doesn't mean that it is necessarily ruling over the world.

It's very interesting to see the manner in which Artificial Intelligence is slowly making the running of businesses to be so much easier. The possibility of solving problems with close to zero error margins results in better running of your business as well as getting answers you require in even shorter periods. The main setback however is that in as much as this concept is very effective in running of businesses, it requires quite a huge capital for it to be installed. This basically mean that small-scale businesses may not find it cost effective to use AI in their companies, locking them out from enjoying AI benefits.

It can be concluded that the main aim of Artificial Intelligence in business is to facilitate the provision of software that has the ability to explain output and reason with connection to input. AI provides interactions that are human-like as well as offering support in making decisions for certain tasks. However, Artificial Intelligence hasn't replaced human beings and it still won't, even in the days to come.